METHODS IN MOLECULAR BIOLOGY

Series Editor
John M. Walker
School of Life and Medical Sciences
University of Hertfordshire
Hatfield, Hertfordshire, UK

For further volumes:
http://www.springer.com/series/7651

For over 35 years, biological scientists have come to rely on the research protocols and methodologies in the critically acclaimed *Methods in Molecular Biology* series. The series was the first to introduce the step-by-step protocols approach that has become the standard in all biomedical protocol publishing. Each protocol is provided in readily-reproducible step-by-step fashion, opening with an introductory overview, a list of the materials and reagents needed to complete the experiment, and followed by a detailed procedure that is supported with a helpful notes section offering tips and tricks of the trade as well as troubleshooting advice. These hallmark features were introduced by series editor Dr. John Walker and constitute the key ingredient in each and every volume of the *Methods in Molecular Biology* series. Tested and trusted, comprehensive and reliable, all protocols from the series are indexed in PubMed.

Live-Cell RNA Imaging

Methods and Protocols

Edited by

Shin-ichi Sato

Faculty of Advanced Science and Technology, Kumamoto University, Kumamoto, Japan

Humana Press

Editor
Shin-ichi Sato
Faculty of Advanced Science and Technology
Kumamoto University
Kumamoto, Japan

ISSN 1064-3745 ISSN 1940-6029 (electronic)
Methods in Molecular Biology
ISBN 978-1-0716-4247-4 ISBN 978-1-0716-4248-1 (eBook)
https://doi.org/10.1007/978-1-0716-4248-1

This Humana imprint is published by the registered company Springer Science+Business Media, LLC, part of Springer Nature.
The registered company address is: 1 New York Plaza, New York, NY 10004, U.S.A.

Preface

Spatiotemporal regulation of gene expression within living cells is controlled by numerous factors, including mRNA stability and transport, and gene transcription. While in situ hybridization has conventionally been employed to determine the intracellular localization of RNAs in fixed cells, the method provides merely a static snapshot captured at the moment of fixation. In contrast, live-cell imaging of RNAs enables the observation of their intracellular distribution and dynamics, facilitating the measurement of spatiotemporal gene expression.

This book is a comprehensive guide to live-cell RNA imaging techniques and methodologies that have been successfully used to study the dynamics or function of various RNAs in cells. The topics covered encompass a selection of robust, state-of-the-art techniques for live-cell RNA imaging, including the detection of aptamer-tagged mRNA molecules utilizing fluorogenic proteins, monitoring endogenous mRNA via a split-protein-based probe within neuronal cells, sensitive detection of low-abundance RNAs employing catalytic hairpin assembly circuits, visualizing the transcriptional repression of target genes using the MS2 system within plant cells, intracellular miRNA imaging with high sensitivity and specificity utilizing DNA-MOF probes, and identification of RNA G-quadruplex structures or nucleolar RNA through fluorescent probes, among other advanced techniques.

These methodologies provide researchers with a diverse toolkit for investigating RNA dynamics and function within live cells, thereby contributing to advancements in our understanding of cellular processes at the molecular level.

Kumamoto, Japan *Shin-ichi Sato*

Contents

Contributors

EDOUARD BERTRAND • *Institut de Génétique Humaine, Univ. Montpellier, CNRS, Montpellier, France; Equipe labélisée Ligue Nationale Contre le Cancer, Montpellier, France*

ZHE CHEN • *Yunnan Key Laboratory of Ecological Protection and Resource Utilization of River-lake Networks, School of Ecology and Environmental Sciences, Institute for Ecological Research and Pollution Control of Plateau Lakes, Yunnan University, Kunming, China*

YICONG DAI • *Beijing Life Science Academy, Beijing, China; Department of Chemistry, Center for BioAnalytical Chemistry, Key Laboratory of Bioorganic Phosphorus Chemistry & Chemical Biology, Tsinghua University, Beijing, China; New Cornerstone Science Laboratory, Shenzhen, China; Center for Bioanalytical Chemistry, Hefei National Laboratory of Physical Sciences at the Microscale, University of Science and Technology of China, Hefei, China*

PASCALE DAVID • *Aix Marseille Univ, CEA, CNRS, BIAM, UMR7265, SAVE (Signalisation pour l'Adaptation des Végétaux à leur Environnement), Saint-Paul, France*

THIERRY DESNOS • *Aix Marseille Univ, CEA, CNRS, BIAM, UMR7265, SAVE (Signalisation pour l'Adaptation des Végétaux à leur Environnement), Saint-Paul, France*

MASATOSHI EGUCHI • *Department of Chemistry, School of Science, The University of Tokyo, Tokyo, Japan*

JIAO-NA HAN • *State Key Laboratory of Chemo/Biosensing and Chemometrics, College of Chemistry and Chemical Engineering, Hunan Provincial Key Laboratory of Biomacromolecular Chemical Biology, Hunan University, Changsha, China*

SAHAR HANI • *Aix Marseille Univ, CEA, CNRS, BIAM, UMR7265, SAVE (Signalisation pour l'Adaptation des Végétaux à leur Environnement), Saint-Paul, France*

HONGWEI HOU • *Beijing Life Science Academy, Beijing, China; Department of Chemistry, Center for BioAnalytical Chemistry, Key Laboratory of Bioorganic Phosphorus Chemistry & Chemical Biology, Tsinghua University, Beijing, China; New Cornerstone Science Laboratory, Shenzhen, China; Center for Bioanalytical Chemistry, Hefei National Laboratory of Physical Sciences at the Microscale, University of Science and Technology of China, Hefei, China*

PING PING HU • *Chongqing Key Laboratory for Pharmaceutical Metabolism Research, College of Pharmacy, Chongqing Medical University, Chongqing, China; Key Laboratory of Biomedical Analytics, Chongqing Science and Technology Bureau, College of Pharmaceutical Sciences, Southwest University, Chongqing, China*

CHENG ZHI HUANG • *Key Laboratory of Biomedical Analytics, Chongqing Science and Technology Bureau, College of Pharmaceutical Sciences, Southwest University, Chongqing, China*

SAMIE R. JAFFREY • *Department of Pharmacology, Weill Cornell Medicine, Cornell University, New York, NY, USA*

JIAN-HUI JIANG • *State Key Laboratory of Chemo/Bio-Sensing and Chemometrics, College of Chemistry and Chemical Engineering, College of Biomedical Sciences, Hunan University, Changsha, China*

ARUNI KARUNANAYAKE MUDIYANSELAGE • *Department of Chemistry, University of Massachusetts, Amherst, MA, USA*

YOUSUKE KATSUDA • *Faculty of Advanced Science and Technology, Kumamoto University, Kumamoto, Japan*
SHI KUANG • *State Key Laboratory of Chemo/Biosensing and Chemometrics, College of Chemistry and Chemical Engineering, Hunan Provincial Key Laboratory of Biomacromolecular Chemical Biology, Hunan University, Changsha, China*
CONGCONG LI • *Cancer Metastasis Alert and Prevention Center, Fujian Provincial Key Laboratory of Cancer Metastasis Chemoprevention and Chemotherapy, State Key Laboratory of Photocatalysis on Energy and Environment and College of Chemistry, Fuzhou University, Fuzhou, China*
JINGHONG LI • *Beijing Life Science Academy, Beijing, China; Department of Chemistry, Center for BioAnalytical Chemistry, Key Laboratory of Bioorganic Phosphorus Chemistry & Chemical Biology, Tsinghua University, Beijing, China; New Cornerstone Science Laboratory, Shenzhen, China; Center for Bioanalytical Chemistry, Hefei National Laboratory of Physical Sciences at the Microscale, University of Science and Technology of China, Hefei, China*
ZHAOHUI LI • *College of Chemistry, Institute of Analytical Chemistry for Life Science, Henan Joint International Research Laboratory of Green Construction of Functional Molecules and Their Bioanalytical Applications, Zhengzhou Key Laboratory of Functional Nanomaterial and Medical Theranostic, Zhengzhou University, Zhengzhou, China*
YONG LIU • *Yunnan Key Laboratory of Ecological Protection and Resource Utilization of River-lake Networks, School of Ecology and Environmental Sciences, Institute for Ecological Research and Pollution Control of Plateau Lakes, Yunnan University, Kunming, China*
XINGYU LUO • *State Key Laboratory of Chemo/Biosensing and Chemometrics, College of Chemistry and Chemical Engineering, Hunan Provincial Key Laboratory of Biomacromolecular Chemical Biology, Hunan University, Changsha, China*
KENNETH R. LYON • *Department of Biochemistry and Molecular Biology, Colorado State University, Fort Collins, CO, USA*
HONG-MIN MENG • *College of Chemistry, Institute of Analytical Chemistry for Life Science, Henan Joint International Research Laboratory of Green Construction of Functional Molecules and Their Bioanalytical Applications, Zhengzhou Key Laboratory of Functional Nanomaterial and Medical Theranostic, Zhengzhou University, Zhengzhou, China*
CAROLINE MERCIER • *Aix Marseille Univ, CEA, CNRS, BIAM, UMR7265, SAVE (Signalisation pour l'Adaptation des Végétaux à leur Environnement), Saint-Paul, France*
TATSUYA MORISAKI • *Department of Biochemistry and Molecular Biology, Colorado State University, Fort Collins, CO, USA*
ZHOU NIE • *State Key Laboratory of Chemo/Biosensing and Chemometrics, College of Chemistry and Chemical Engineering, Hunan Provincial Key Laboratory of Biomacromolecular Chemical Biology, Hunan University, Changsha, China*
SEIICHI NISHIZAWA • *Department of Chemistry, Graduate School of Science, Tohoku University, Sendai, Japan*
LAURENT NUSSAUME • *Aix Marseille Univ, CEA, CNRS, BIAM, UMR7265, SAVE (Signalisation pour l'Adaptation des Végétaux à leur Environnement), Saint-Paul, France*
TAKEAKI OZAWA • *Department of Chemistry, School of Science, The University of Tokyo, Tokyo, Japan*
TAKASHI SAKAMOTO • *Faculty of Systems Engineering, Graduate School of Systems Engineering, Wakayama University, Wakayama, Japan*
SHIN-ICHI SATO • *Faculty of Advanced Science and Technology, Kumamoto University, Kumamoto, Japan*

YUSUKE SATO • *Department of Chemistry, Graduate School of Science, Tohoku University, Sendai, Japan*
TIMOTHY J. STASEVICH • *Department of Biochemistry and Molecular Biology, Colorado State University, Fort Collins, CO, USA*
MICHIYUKI SUZUKI • *Department of Chemistry, Graduate School of Science, Tohoku University, Sendai, Japan*
XUCONG TENG • *Beijing Life Science Academy, Beijing, China; Department of Chemistry, Center for BioAnalytical Chemistry, Key Laboratory of Bioorganic Phosphorus Chemistry & Chemical Biology, Tsinghua University, Beijing, China; New Cornerstone Science Laboratory, Shenzhen, China; Center for Bioanalytical Chemistry, Hefei National Laboratory of Physical Sciences at the Microscale, University of Science and Technology of China, Hefei, China*
XING WANG • *College of Chemistry, Institute of Analytical Chemistry for Life Science, Henan Joint International Research Laboratory of Green Construction of Functional Molecules and Their Bioanalytical Applications, Zhengzhou Key Laboratory of Functional Nanomaterial and Medical Theranostic, Zhengzhou University, Zhengzhou, China*
JIAHUI WU • *Department of Pharmacology, Weill Cornell Medicine, Cornell University, New York, NY, USA; Department of Chemistry, University of Massachusetts, Amherst, MA, USA*
ZAI-SHENG WU • *Cancer Metastasis Alert and Prevention Center, Fujian Provincial Key Laboratory of Cancer Metastasis Chemoprevention and Chemotherapy, State Key Laboratory of Photocatalysis on Energy and Environment and College of Chemistry, Fuzhou University, Fuzhou, China; Key Laboratory of Laboratory Medicine, Ministry of Education of China, and Zhejiang Provincial Key Laboratory of Medical Genetics, School of Laboratory Medicine and Life Sciences, Wenzhou Medical University, Wenzhou, China*
ZHENKUN WU • *State Key Laboratory of Chemo/Bio-Sensing and Chemometrics, College of Chemistry and Chemical Engineering, College of Biomedical Sciences, Hunan University, Changsha, China*
KENJI YATSUZUKA • *Department of Biomolecular Engineering, Graduate School of Engineering, Nagoya University, Nagoya, Japan*
HIDEAKI YOSHIMURA • *Department of Chemistry, School of Science, The University of Tokyo, Tokyo, Japan*
MINGXU YOU • *Department of Chemistry, University of Massachusetts, Amherst, MA, USA*
LEI ZHAN • *Key Laboratory of Biomedical Analytics, Chongqing Science and Technology Bureau, College of Pharmaceutical Sciences, Southwest University, Chongqing, China*
HONGYAN ZHANG • *Department of Chemistry, Tianjin Key Laboratory of Molecular Optoelectronic Sciences, School of Science, Tianjin University, Tianjin, People's Republic of China*
JIAHENG ZHANG • *State Key Laboratory of Chemo/Biosensing and Chemometrics, College of Chemistry and Chemical Engineering, Hunan Provincial Key Laboratory of Biomacromolecular Chemical Biology, Hunan University, Changsha, China*
LIBING ZHANG • *Department of Chemistry, Tianjin Key Laboratory of Molecular Optoelectronic Sciences, School of Science, Tianjin University, Tianjin, People's Republic of China*
SU-JING ZHAO • *State Key Laboratory of Chemo/Bio-Sensing and Chemometrics, College of Chemistry and Chemical Engineering, College of Biomedical Sciences, Hunan University, Changsha, China*
LIN LING ZHENG • *Biomedical Analysis Center, Army Medical University, Chongqing, China*

Chapter 1

Live-Cell Imaging of mRNA Using a Pepper RNA Tag

Jiahui Wu and Samie R. Jaffrey

Abstract

Live-cell imaging of mRNA enables tracking of mRNA localization and its dynamics in real time. This is fundamentally important in understanding how cells use RNA to regulate gene expression and orchestrate biological processes. Here, we describe a method of using an engineered RNA tag, called Pepper RNA tag, to visualizing mRNA in living cells. In this method, an mRNA of interest engineered to contain the Pepper RNA tag turns on the fluorescence signals of fluorogenic proteins, which enables tracking of mRNA in living cells.

Key words Live-cell RNA imaging, Pepper RNA tag, Fluorogenic proteins

1 Introduction

mRNAs are universal messenger molecules, which carry genetic information from the genome to different parts of the cells for making functional proteins. The localization of mRNAs is crucial for cells to maintain their functions and health. This is because the exact positioning of mRNAs in the cell determines where the functional proteins, encoded by these mRNAs, are made. Indeed, intracellular trafficking of mRNAs plays a key role in modulating asymmetric protein distribution linked to embryonic patterning, asymmetric cell division, and cell migration [1]. Furthermore, defective mRNA localization leads to the loss of normal cell polarity, which is linked to diseases, such as musculodegenerative or neurodegenerative diseases [2]. Therefore, being able to image and track the dynamics of mRNAs in living cells can provide critical insights into understanding the pathogenesis of diseases and to guide therapeutic interventions.

Imaging mRNAs is a highly challenging task because mRNA is in low abundance in cells and conferring fluorescence signals to mRNA is difficult. To address this problem, researchers have developed two major methods for imaging mRNAs in living cells:

Shin-ichi Sato (ed.), *Live-Cell RNA Imaging: Methods and Protocols*, Methods in Molecular Biology, vol. 2875,
https://doi.org/10.1007/978-1-0716-4248-1_1,

(1) tethering fluorescent proteins to the mRNAs of interest using RNA hairpins derived from phages [3, 4] and (2) selectively turning on the fluorescence signals of otherwise non-fluorescent dyes on the mRNAs of interest using fluorogenic aptamers [5–12]. Tethering methods are problematic because the untethered fluorescent protein exhibits fluorescence. Thus, it can be challenging to detect the fluorescent protein-bound mRNA from the unbound fluorescent protein. Fluorogenic aptamers can be problematic since they often do not have as robust fluorescence or photostability as many fluorescent proteins.

Here, we demonstrate the application of the third major method for imaging mRNAs in living cells using Pepper RNA-regulated fluorogenic proteins [13]. This method relies on tethering fluorescent proteins to an mRNA, but also exhibits selective fluorescence when the fluorescent protein interacts with the mRNA. Thus, unbound fluorescent protein does not interfere with detection of the mRNA to the same degree as previous tethering methods. In this method, a destabilization domain, called tDeg, is added to the C-terminus of a fluorescent protein. This causes any unbound fluorescent protein to be degraded rapidly in the cell. However, this tDeg-tagged fluorescent protein is stabilized and not degraded only when it is bound to a specific RNA aptamer, called Pepper. Thus, the unbound fluorescent protein has low fluorescence signal since it is rapidly degraded, thus making it easier to visualize the mRNA that contains the bound fluorescent protein. By fusing multiple Pepper aptamers to the 3′untranslated region of the mRNA of interest, a Pepper-tagged mRNA can be selectively imaged in cells with markedly lower background than existing fluorescent protein tethering methods (Fig. 1).

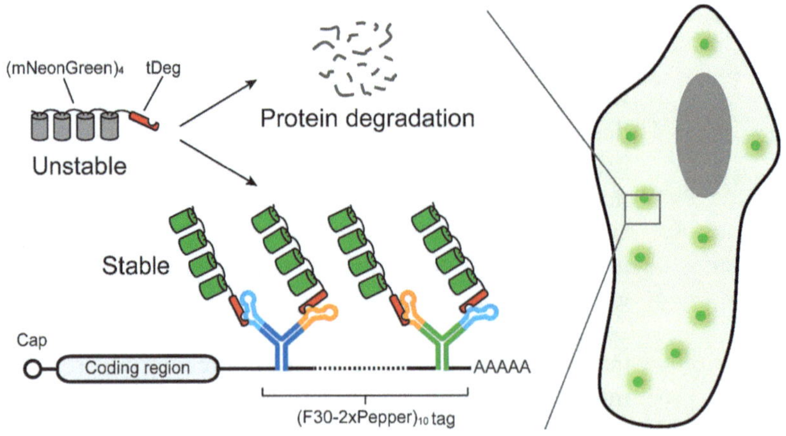

Fig. 1 Imaging an mRNA of interest using the Pepper RNA tag and fluorogenic proteins. In this method, a destabilization domain, called tDeg, is added to the C-terminus of four concatenated fluorescent proteins, $(mNeonGreen)_4$. This causes $(mNeonGreen)_4$-tDeg to be degraded rapidly in the cell. However, $(mNeon-Green)_4$-tDeg can only be stabilized when it binds to an RNA aptamer, called Pepper. In this way, by fusing Pepper RNA tag ($(F30\text{-}2xPepper)_{10}$) to the mRNA of interest, this Pepper-tagged mRNA can selectively turn on the fluorescence signals of fluorescent protein for mRNA imaging in living cells

2 Materials

2.1 Cell Culture

1. U2OS cells.
2. Cell culture medium: DMEM supplemented with 10% fetal bovine serum, 100 U/mL penicillin and 100 μg/mL of streptomycin.
3. T75 cell culture flask.
4. 1× PBS (1× phosphate-buffered saline), premade by manufacturer.
5. 1× TrypLE Express (no phenol red), premade by manufacturer.
6. 15 mL tube.
7. 50 mL tube.
8. Trypan blue solution, 0.4%, premade by manufacturer.
9. 35 mm imaging dish with glass bottom, precoated with poly-D-lysin by manufacturer (*see* **Note 1**).

2.2 Transfection

1. Opti-MEM, premade by manufacturer (*see* **Note 2**).
2. DNA plasmid: CMV-mCherry-$(F30\text{-}2xPepper)_{10}$ (Addgene plasmid number: #129401) (*see* **Notes 3** and **4**).
3. DNA plasmid: miniCMV-$(mNeonGreen)_4$-tDeg (Addgene plasmid number: #129402) (*see* **Notes 3** and **4**).
4. FuGENE® HD transfection reagent, premade by manufacturer.

2.3 Imaging

1. 100× GlutaMAX™, premade by manufacturer.
2. 100 mM sodium pyruvate, premade by manufacturer.
3. Imaging medium: Phenol red-free DMEM supplemented with 10% fetal bovine serum, 100 U/mL penicillin and 100 μg/mL of streptomycin, 1× GlutaMAX™, and 1 mM sodium pyruvate.
4. Hoechst dye solution, premade by manufacturer.
5. Immersion oil (N = 1.520), premade by manufacturer.

3 Methods

3.1 Culturing and Seeding Cells

1. Maintain U2OS cells with cell culture medium in a T75 cell culture flask at 37 °C with 5% CO_2 until cell confluency reaches ~80%.
2. Set water bath temperature to 37 °C, and then prewarm cell culture medium, 1× PBS, and 1× TrypLE Express in water bath to reach 37 °C.

3. Aspirate cell culture medium from the T75 cell culture flask with U2OS cells, then rinse the U2OS cells with ~10 mL prewarmed 1× PBS once, and aspirate the 1× PBS.
4. After aspirating the 1× PBS, add 2 mL 1× TrypLE Express to the same T75 cell culture flask with U2OS cells, and then place the flask back to the cell incubator at 37 °C with 5% CO_2 for 5 to 10 min to let cells detach from the flask.
5. After incubation, take the T75 cell culture flask out of the incubator, and add ~8 mL of prewarmed cell culture medium to the flask. Gently pipette the 1× TrypLE Express and cell culture medium mixture using a serological pipette to detach cells from the flask.
6. Then transfer the detached cells into a 15 mL tube, and centrifuge this 15 mL tube at 300 g for 3 min.
7. After centrifugation, cells should be pelleted at the bottom of the 15 mL tube. Aspirate supernatant carefully without disrupting the cell pellet.
8. Resuspend cells using ~15 mL prewarmed cell culture medium. Pipette gently to ensure cells are fully resuspended, and then transfer the resuspended cells in a 50 mL tube.
9. Mix 20 μL of resuspend cells with 20 μL of trypan blue solution, then pipette 10 μL of this mixture to a hemocytometer, and count the number of cells.
10. Based on the cell density calculated from **step 9**, dilute U2OS cells to a density of 1×10^5 per 1 mL using cell culture medium. Then, seed 2×10^5 U2OS cells in 2 mL of cell culture medium to each 35 mm imaging dish (coated with poly-D-lysin), shake gently to ensure cells are evenly distributed in the dish, and put dishes back to the incubator at 37 °C with 5% CO_2 and culture overnight (*see* **Note 5**).

3.2 Transfection

1. Transfect U2OS cells 18–24 h after seeding them on the 35 mm imaging dish. To transfect one 35 mm imaging dish of U2OS cells, mix the following reagents in a 1.7 mL tube: 125 μL of Opti-MEM, 1.4 μg of CMV-mCherry-(F30-2xPepper)$_{10}$ DNA plasmids, 1.4 μg of miniCMV-(mNeonGreen)$_4$-tDeg DNA plasmids, and 9.6 μL of FuGENE® HD reagent (*see* **Note 6**).
2. Gently pipette to mix the above mixture thoroughly, and incubate at room temperature for 15 min.
3. Pipette dropwise the above transfection mixture to U2OS cells in a 35 mm imaging dish. Gently shake the dish to ensure transfection mixture is well-mixed with the cell culture medium. Transfer the 35 mm imaging dish back to the incubator at 37 °C with 5% CO_2, and culture overnight.

4. On the next day, prewarm 1× PBS and cell culture medium in water bath to 37 °C.
5. Aspirate cell culture medium from 35 mm imaging dish about 18–24 h after transfection. Gently rinse the transfected U2OS cells with 2 mL prewarmed 1× PBS. Aspirate 1× PBS, and then add 2 mL fresh prewarmed cell culture medium to U2OS cells. Transfer the 35 mm imaging dish back to incubator at 37 °C with 5% CO_2, and culture overnight.

3.3 Imaging

1. Prewarmed 1× PBS and imaging medium in water bath to 37 °C.
2. Prewarmed microscope imaging stage and environmental chamber to 37 °C.
3. Aspirate cell culture medium from 35 mm imaging dish. Gently rinse the transfected U2OS cells with 2 mL prewarmed 1× PBS. Aspirate 1× PBS, and then add 2 mL fresh prewarmed imaging medium supplemented with Hoechst dye to U2OS cells. Then incubate U2OS cells at 37 °C with 5% CO_2 for 5 min before imaging.
4. Add sufficient immersion oil to a 100×/1.4-NA oil objective, and then transfer U2OS cells in 35 mm imaging dish to prewarmed microscope imaging stage. Then, use phase illumination, and adjust the height of the objective to focus on adhered U2OS cells.
5. Switch imaging mode from phase to fluorescence acquisition. Use a TRITC filter cube (with excitation filter 542 ± 13.5 nm, dichroic mirror with a reflection band of 547–565 nm and a transmission band of 576–630 nm, and emission filter 594 ± 22.5 nm) to identify transfected U2OS cells expressing the red fluorescent protein, mCherry (*see* **Note** 7).
6. Switch to a YFP filter cube (with excitation filter 500 ± 12 nm, dichroic mirror 520 nm (long pass) and emission filter 542 ± 13.5 nm) to image reporter mRNA, mCherry-$(F30\text{-}2xPepper)_{10}$. Adjust exposure time in the microscope software to acquire fluorescence images (*see* **Note 8**).

4 Notes

1. For cell types that are easy to adhere to glass surface, such as U2OS cells, no extra coating is needed normally. For cell types that are more prone to detach from glass surface, consider coating the 35 mm imaging dish with mouse laminin I.
2. For transfection using FuGENE® HD, Opti-MEM can be replaced by double-distilled water.

3. DNA plasmids can be purified using silica membrane-based miniprep columns. However, if high cytotoxicity is observed after transfection, consider using anion exchange-based maxiprep columns to minimize the levels of endotoxin from bacteria used for propagating the DNA plasmids.
4. In this method, we recommend using the $(F30\text{-}2xPepper)_{10}$ tag, which contains ten repeats of 2xPepper within an F30 RNA folding scaffold [14] (a total number of 20 Peppers). In a previous study, we have tested 20 Peppers without the F30 RNA folding scaffold and 10 Peppers with and without the F30 RNA folding scaffold, and we found that $(F30\text{-}2xPepper)_{10}$ tag gives the highest fluorescence signals in the Pepper-tagged mRNA [13]. To maximize the fluorescence signals of the Pepper-tagged mRNA, we use $(mNeonGreen)_4$-tDeg, which contains four concatenated mNeonGreens fused to tDeg. In a previous study, we found that this gives the highest signals on the Pepper-tagged mRNA compared to one, two, or three copies of mNeonGreen [13].
5. Adjust the number of cells seeded to each 35 mm imaging dish according to cell type. We typically seed 2×10^5 U2OS cells in 2 mL of cell culture medium per 35 mm imaging dish. However, the optimal number of cells seeded to each 35 mm imaging dish should be optimized accordingly to each experimental need.
6. An optimal amount of $(mNeonGreen)_4$-tDeg plasmid will give the highest mRNA fluorescence signal-to-background fluorescence signal ratio in fluorescence imaging. Normally, we use 1.4 μg of miniCMV-$(mNeonGreen)_4$-tDeg in the transfection step. If the mRNA fluorescence signals miniCMV-$(mNeonGreen)_4$-tDeg are too low, consider using UbC-$(mNeonGreen)_4$-tDeg as the fluorogenic protein. A starting point could be using 0.28 μg of UbC-$(mNeonGreen)_4$-tDeg (Addgene plasmid #129404) with 1.12 μg of pUC19 (as a diluent DNA) as the fluorogenic protein in the transfection step. However, the amount of plasmid encoding $(mNeonGreen)_4$-tDeg should be titrated and optimized case by case. Using too high of an amount of UbC-$(mNeonGreen)_4$-tDeg will lead to background fluorescence from overexpression of $(mNeonGreen)_4$-tDeg.
7. A general rule of thumb for fluorescence imaging is to avoid prolonged excitation to the specimen to avoid photobleaching of the fluorophores.
8. It is important to adjust exposure time to ensure the highest possible mRNA fluorescence signal without any pixel saturation. To determine the proper exposure time, acquire images with an exposure time between 50 ms and 500 ms. If the

fluorescence signals from the Pepper-tagged mRNA are too low, consider taking out all the neutral-density filter from the excitation light source to maximize excitation light intensity. In addition, the fluorescence signals of single mRNA molecules are relatively low. We recommend using a sensitive camera, such as an EMCCD camera or a CMOS camera, to capture images or acquiring movies.

Acknowledgments

This work was supported by NIH grant R35 NS111631 (SRJ), Starr Cancer Consortium Grant I12-0051 (SRJ), and UMass start-up funds (JW).

References

1. Cody NAL, Iampietro C, Lécuyer E (2013) The many functions of mRNA localization during normal development and disease: from pillar to post. Wiley Interdiscip Rev Dev Biol 2: 781–796
2. Liu-Yesucevitz L, Bassell GJ, Gitler AD et al (2011) Local RNA translation at the synapse and in disease. J Neurosci 31:16086–16093
3. Bertrand E, Chartrand P, Schaefer M et al (1998) Localization of ASH1 mRNA particles in living yeast. Mol Cell 2:437–445
4. Chao JA, Patskovsky Y, Almo SC, Singer RH (2008) Structural basis for the coevolution of a viral RNA-protein complex. Nat Struct Mol Biol 15:103–105
5. Paige JS, Wu KY, Jaffrey SR (2011) RNA mimics of green fluorescent protein. Science 333:642–646
6. Filonov GS, Moon JD, Svensen N, Jaffrey SR (2014) Broccoli: rapid selection of an RNA mimic of green fluorescent protein by fluorescence-based selection and directed evolution. J Am Chem Soc 136:16299–16308
7. Song W, Filonov GS, Kim H et al (2017) Imaging RNA polymerase III transcription using a photostable RNA-fluorophore complex. Nat Chem Biol 13:1187–1194
8. Braselmann E, Wierzba AJ, Polaski JT et al (2018) A multicolor riboswitch-based platform for imaging of RNA in live mammalian cells. Nat Chem Biol 14:964–971
9. Chen X, Zhang D, Su N et al (2019) Visualizing RNA dynamics in live cells with bright and stable fluorescent RNAs. Nat Biotechnol 37: 1287–1293
10. Bouhedda F, Fam KT, Collot M et al (2020) A dimerization-based fluorogenic dye-aptamer module for RNA imaging in live cells. Nat Chem Biol 16:69–76
11. Cawte AD, Unrau PJ, Rueda DS (2020) Live cell imaging of single RNA molecules with fluorogenic Mango II arrays. Nat Commun 11:1283
12. Dey SK, Filonov GS, Olarerin-George AO et al (2021) Repurposing an adenine riboswitch into a fluorogenic imaging and sensing tag. Nat Chem Biol. https://doi.org/10.1038/s41589-021-00925-0
13. Wu J, Zaccara S, Khuperkar D et al (2019) Live imaging of mRNA using RNA-stabilized fluorogenic proteins. Nat Methods 16:862–865
14. Filonov GS, Kam CW, Song W, Jaffrey SR (2015) In-gel imaging of RNA processing using broccoli reveals optimal aptamer expression strategies. Chem Biol 22:649–660

Chapter 2

A Method for Bioluminescence-Based RNA Monitoring Using Split-Luciferase Reconstitution Techniques

Masatoshi Eguchi, Hideaki Yoshimura, and Takeaki Ozawa

Abstract

A bioluminescence-based RNA monitoring method in living cells was developed using a split NanoLuc (NLuc) reconstitution technique. For specific recognition of a target RNA sequence, a mutant PUM-HD (mPUM) was used. The method was applied to β-actin mRNA in various cells, including primary cultures of rat hippocampal neurons. It allowed for continuous observation of intracellular localization distribution of the target mRNA in living cells, providing insights into various biological phenomena involving intracellular RNA localization and dynamics.

Key words RNA, Live-cell imaging, Luciferase, Reconstitution, Bioluminescence

1 Introduction

Fundamental roles of intracellular RNA location and motion in biological phenomena have been widely recognized [1–4]. One of the RNAs well studied in its roles is β-actin mRNA. The intracellular location of β-actin mRNA is essential for cell spreading, migration, and axonal outgrowth [5, 6]. For a detailed understanding of these phenomena, methods are required to monitor the time course of RNA localization and distribution through live-cell RNA imaging. Moreover, RNAs are a direct target molecule in various gene expression analysis techniques. Technological advances in various comprehensive gene expression analyses have promoted the discovery of genes associated with a variety of physiological events by identifying and quantifying the corresponding RNAs [7]. However, these methods fail to obtain spatiotemporal information on RNAs because of destructive approaches. This emphasizes the need for non-destructive live-cell imaging techniques for spatiotemporal analysis of RNAs.

RNA imaging technologies applicable to living cells for spatiotemporal analysis of the target RNA distribution and motion can be

Shin-ichi Sato (ed.), *Live-Cell RNA Imaging: Methods and Protocols*, Methods in Molecular Biology, vol. 2875, https://doi.org/10.1007/978-1-0716-4248-1_2,

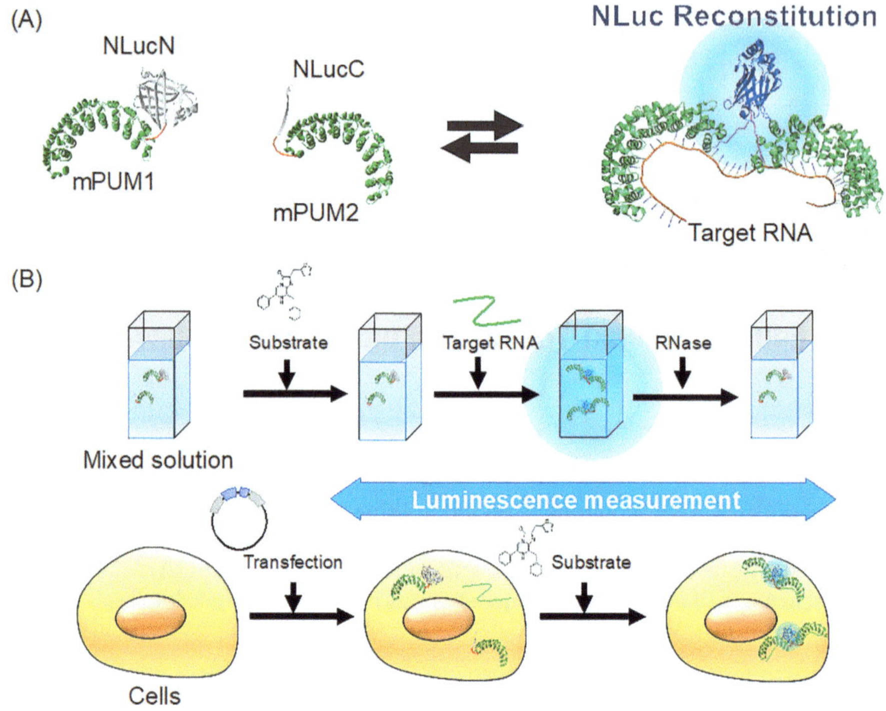

Fig. 1 Detection and visualization of target RNA using bioluminescent probes. (**a**) A schematic of the RNA recognition mechanism of the bioluminescent probes. (**b**) A schematic of the bioluminescent probe protein expressed in a test tube and within cells demonstrates the mechanism

achieved by establishing two core techniques: a selective recognition of a target RNA and a visualization of the labeled RNA. An RNA-binding protein domain, PUM-HD, is a useful candidate for the tool for selective RNA labeling [8–13]. PUM-HD is an RNA-binding domain of human PUMILIO1. The crystal structure shows that PUM-HD consists of eight repeating motifs, each forming specific interactions with a single RNA base, resulting in selective recognition of 8-based RNA in total [14]. It is a noteworthy point of PUM-HD for an RNA recognition tool that PUM-HD mutants (mPUM) can be designed tailor-made to recognize a particular 8-based RNA sequence based on the information of the crystal structure (Fig. 1a) [13, 15–17]. A fusion protein consisting of a fluorescent protein and mPUM(s) designed to recognize a portion in a target RNA of interest can be used as an RNA probe for monitoring the dynamics of the target RNA in living cells [9]. In addition, the split fluorescent protein reconstitution technique using mPUM-based RNA probes suppresses background fluorescence from excess probes and provides more precise RNA imaging [8, 12]. To monitor the target RNA labeled with the probe in living cells, fluorescence microscopy has been the primary approach for RNA visualization. While having advantages in sensitivity and selectivity in live-cell observations, fluorescence imaging also has drawbacks, such as photobleaching and phototoxicity. A bioluminescence imaging approach offers a potential solution to the drawbacks of fluorescence imaging [18]. The bioluminescence approach, which uses luciferase-catalyzed substrate oxidation to produce luminescence, has recently attracted attention and interest as a unique imaging technique [19, 20]. Bioluminescence imaging, which does not require an excitation light, is ideal for long-term live-cell imaging due to its low phototoxicity and high signal-to-background ratio. Luciferases can be applied for the split-reconstitution technique, as is the case of fluorescent protein, and the reconstitution process is reversible, which suggests the potential of the luciferase reconstitution technique to apply quantitative analysis of the target molecule.

The protocol in this chapter introduces a new method to monitor the intracellular localization and dynamics of target RNAs using a bioluminescence-based RNA probe (Fig. 1b) [21]. In the study on which this protocol was generated, we developed a bioluminescence-based RNA probe for live-cell imaging, enabling tracking of RNA localization and dynamics during various cellular physiological events. We adopted split NanoLuc (NLuc) reconstitution and mPUM-mediated selective RNA recognition [14, 17, 22, 23]. We selected mouse β-actin mRNA as a proof-of-concept target because of its well-investigated localization properties and dynamics in living cells (Fig. 2a). Using the bioluminescent probe, the spatiotemporal distribution of the target RNA was visualized in living cells over time. In addition, this method was applied to primary cultures of rat hippocampal neurons, highlighting its potential applications in neuroscience research (Fig. 2b).

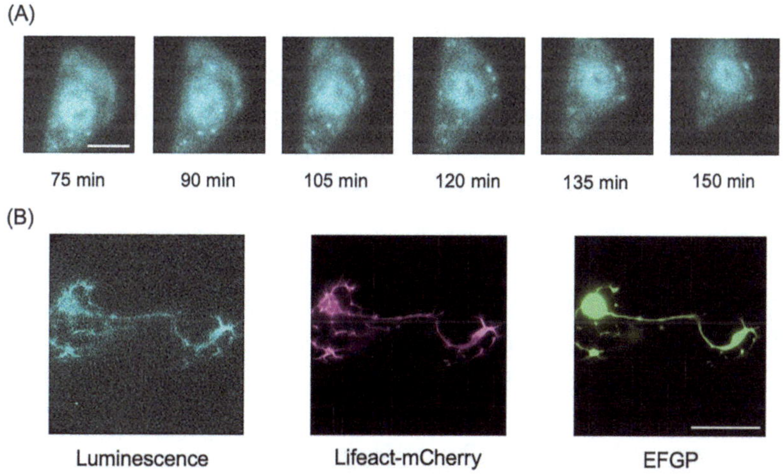

Fig. 2 Live imaging of cells expressing the bioluminescent probe. (**a**) Time-lapse luminescence imaging of cultured cells expressing the bioluminescent probe, tracking the localization of RNA. (**b**) Microscopic images showing luminescence and fluorescence imaging in neurons co-expressing the bioluminescent probe, Lifeact-mCherry, and EGFP

2 Materials

2.1 Plasmids

1. The pcDNA4/V5-His ver. B vector incorporating the NLucN-mPUM1 gene, which consists of the genes of the NLucN fragment (1–156 a.a, LgBiT) and the RNA binding domains of a human PUMILIO1 (PUM-HD) (828–1176 a.a) mutant containing specific mutations at positions C935S and Q939E [8, 15]. The two genes encoding the NLucN and the PUM-HD mutant are linked together by a GS linker (*see* **Note 1**).
2. The pcDNA4/V5-His ver. B vector incorporating the gene of mPUM2-NlucC, which consists of the NLucC fragment (SmBit) and PUM-HD with specific mutations at positions S863N, N899S, Q903E, C935N, Q939E, C1007S, and Q1011E [8, 15]. These components are linked together by a GS linker (*see* **Note 2**).
3. pCold I DNA vector incorporating NLucN-mPUM1.
4. pCold I DNA vector incorporating mPUM2-NLucC.
5. pCAG vector incorporating NLucN-mPUM1 and mPUM2-NLucC, which are connected through a bicistronic system involving 2A sequences (*see* **Note 3**).
6. pCAG vector incorporating EGFP.
7. pCAG vector incorporating Lifeact-mCherry.

2.2 Cell Culture and Transfection

1. Dulbecco's modified Eagle medium, DMEM (4.5 g/L glucose) with L-Gln, without sodium pyruvate.
2. Fetal bovine serum (FBS).
3. 35 mM glass-bottom dishes.
4. Poly-L-lysine hydrochloride.
5. Trypsin–EDTA (0.5 g/L-Trypsin/0.53 mmoL/L-EDTA solution with phenol red).
6. TransIT-LT1.
7. Cell counter.

2.3 Luminescence Measurement Using a 96-Well Plate

1. Tristar Multimode Reader LB941.
2. Nano-Glo Luciferase Assay System.
3. Hanks' balanced salt solution (HBSS).

2.4 Probe Expression in E. coli

1. *E. coli* BL21 (DE3) strain.
2. Isopropyl-β-D(-)-thiogalactopyranoside (IPTG).

2.5 Probe Protein Purification

1. ÄKTA purifier system.
2. Binding buffer (20 mM phosphate buffer, pH 7.2, 250 mM NaCl, 20 mM imidazole).
3. Protease inhibitor (complete EDTA-free).
4. 0.22 μm syringe filter.
5. HisTrap FF column.
6. Elution buffer (20 mM phosphate buffer, pH 7.2, 250 mM NaCl, 1 M imidazole).
7. Dilution buffer (20 mM phosphate buffer, pH 7.2).
8. HiTrap Q HP column.
9. Buffer A (20 mM phosphate buffer, pH 7.2).
10. Buffer B (20 mM phosphate buffer, pH 7.2, 1 M NaCl).
11. BCA Protein Assay Kit.

2.6 Time Course Measurement of Bioluminescence Spectra

1. LumiFL SpectroCapture.
2. Nano-Glo Luciferase Assay System.
3. Phosphate-buffered saline (PBS (-)).

2.7 SDS-Page

1. RIPA buffer (Tris–HCl (pH 7.4) 50 mM, NaCl 150 mM, NaF 10 mM, EDTA 1 mM, NP-40 0.5% (v/v), SDS 0.1% (w/v), sodium deoxycholate 0.5% (w/v)).
2. Sample buffer (Tris–HCl pH 6.8, 250 mM, Tris–HCl pH 6.8, 100 mM, SDS 10% (w/v), sucrose 30% (w/v), 2-mercaptoethanol 25%, bromophenol blue 0.02% (w/v)).
3. Coomassie brilliant blue (CBB).
4. Image analyzer.

2.8 Antibodies

1. FLAG-tagged probes were detected by monoclonal anti-FLAG M2 antibody (F3165, 1:2000).
2. His-tagged probes were detected by polyclonal rabbit anti-His-tag antibody (PM032, 1:2500).
3. V5-tagged were detected by V5 Tag Monoclonal Antibody (R960-25, 1:5000).
4. β-actin were detected by Anti-β-actin antibody, mouse monoclonal (A1978, 1:2000).
5. Anti-Mouse IgG, HRP-Linked Whole Ab Sheep (NA931V, 1: 5000).
6. Anti Rabbit IgG, HRP Linked Whole Ab Donkey (NA934V, 1:5000).

2.9 Western Blotting

1. Nitrocellulose membrane.
2. Blocking buffer: 5% skim milk in Tris-buffered saline with Tween 20 (TBST).
3. ECL Select Western blot detection reagents.
4. Image analyzer.

2.10 Live Cell Imaging of PUM-NLuc

1. Nano-Glo Live Cell Substrate.
2. Hanks' balanced salt solution.
3. PDGF-BB.
4. Inverted microscope equipped with a PlanApo N 60×/1.42NA objective lens.
5. EM-CCD camera.

2.11 Primary Cultures of Rat Hippocampal Neurons

1. Preparation medium (Papain, DNase, Hibernate-A medium) (*see* **Note 4**).
2. Glass-bottom dishes subjected to surface plasma treatment and poly-L-lysine coating.
3. Plating medium (Neurobasal-A medium, GlutaMAX, B27, Horse serum, HEPES).
4. Growth medium (Neurobasal-A medium, GlutaMAX, B27).
5. Feeding medium (Neurobasal-A medium, cytosine arabinoside, GlutaMAX, B27).
6. Preparation medium (Papain, DNase, Hibernate-A medium).
7. 4D-Nucleofector.
8. Lipofectamine 2000.

3 Methods

3.1 Cell Cultivation and Transfection

1. Prepare the cultivation medium using DMEM supplemented with 10% FBS and the following mammalian cell lines: mouse embryonic fibroblast NIH3T3, human embryonic kidney 293 (HEK293T), human cervical carcinoma HeLa, and human osteosarcoma U2OS.
2. Culture the cells in a 5% CO_2 atmosphere at 37 °C, ensuring to passage them every 2 or 3 days or when the cell confluency becomes around 80%.
3. Detach the cells with trypsin–EDTA treatment.
4. Count the cell number, and transfer the cell suspension of 2×10^5 cells/mL to a 96-well plate (for luminescence measurement) or poly-L-lysine-coated glass-bottom dishes (for live imaging) (*see* **Notes 5** and **6**).

5. Incubate the cells for at least 24 h at 37 °C and 5% CO_2 in a humidified incubator.
6. Transfect plasmids into the cells using a transfection reagent.

3.2 Luminescence Measurement Using a 96-Well Plate

1. After 24 h from the transfection with the probe expression plasmids, count the cells using a cell counter.
2. Take a 20 μL sample from the suspension of transfected cells, mix it with trypan blue for cell counting, and calculate the cell density of the suspension.
3. Adjust the cell density to 2×10^5 cells/mL.
4. Add 100 μL of this adjusted cell suspension into each well of a 96-well plate to achieve a density of 2×10^4 cells per well (*see* **Note 7**).
5. Prepare a diluted substrate of the Nano-Glo Luciferase Assay System with HBSS (*see* **Note 8**).
6. Add Nano-Glo substrate solution into each well of the 96-well plate to carry out luminescence measurements.
7. Perform luminescence measurements 24 h after plating using a microplate reader.

3.3 Probe Expression in E. coli and Purification

1. Transform the *E. coli* BL21 (DE3) strain with the expression plasmid of the probe in the pCold vector, and cultivate on an LB-Agar plate.
2. Pick and transfer a transformed *E. coli* colony, and incubate in LB medium at 37 °C.
3. Monitor the OD_{600} of the culture medium until it reaches approximately 0.6.
4. Transfer the culture medium to a 15 °C incubator for 30 min to induce a cold shock.
5. Add IPTG to a final concentration of 0.1 mM after the cold shock.
6. Incubate the culture for 24 h at 15 °C to induce protein expression.
7. Centrifuge the culture medium to pellet *E. coli*.
8. *E. coli* pellets were resuspended in a binding buffer supplemented with a protease inhibitor.
9. Sonication was performed for a total of 10 min. Each cycle consisted of 40 s of sonication followed by a 20-s rest. Keep the sample on ice during the sonication.
10. Isolate and filter the resulting supernatant using a 0.22 μm filter.

11. Using the ÄKTA purifier system, apply the filtrated supernatant to a HisTrap FF column equilibrated with the binding buffer (*see* **Note 8**). Subsequently, wash the column with 10 column volumes of binding buffer.
12. Elute the protein adsorbed on the column with an elution buffer.
13. Dilute the eluted fraction containing the probe with dilution buffer to reduce the salt concentration, and filtrate the solution with a 0.22 μm filter.
14. Load the solution to a HiTrap Q HP column equilibrated with Buffer A using the ÄKTA purifier system.
15. Wash the column with Buffer A, and then gradually increase the content of Buffer B to elute the adsorbed proteins.
16. Collect the elution fractions that contain the probe protein, monitoring the absorbance at 280 nM.
17. Load the sample onto a 12% SDS-polyacrylamide gel, and conduct electrophoresis.
18. Stain the gel with CBB and monitor using an image analyzer.

3.4 Luminescence Measurement of Purified Protein Using a Plate Reader

1. Measure the concentration of the purified protein using the BCA method.
2. Dilute and mix the protein solution in PBS to achieve a concentration of 250 μM in a final volume of 100 μL. Add the protein solution to 90 μL with the luminescent substrate in PBS.
3. Add a solution containing RNA to prepare a 10 μL mixture to the protein solution, resulting in a final volume of 100 μL.
4. Mix the solution per well by pipetting.
5. Conduct luminescence measurement of the solution using a luminometer.

3.5 Time Course Measurement of Bioluminescence Spectra

1. Adjust the samples following the method described in Subheading 3.4.
2. Conduct protein concentration measurement.
3. Mix purified protein with the luminescent substrate in PBS.
4. Load the mixture into the luminescence spectrometer, and initiate the measurement.
5. Add RNA through the measurement and continue monitoring.
6. Add RNase through the measurement and continue monitoring.

3.6 Western Blotting

1. For each sample, seed 4×10^5 HEK293T cells on a 6-well plate, and incubate for 24 h.
2. Harvest HEK293T cells 24 h post-transfection by removing the medium and washing once with PBS.
3. Add 100–200 μL of RIPA lysis buffer to each well. Use a cell scraper to detach the cells at 4 °C.
4. Collect the cell lysate in a 1.5 mL tube. Sonication was performed for a total of 5 min. Each cycle consisted of 20 s of sonication followed by a 10-s rest and incubation on ice for 10 min at 4 °C.
5. Centrifuge at 15,000 rpm. For 10 min. at 4 °C.
6. Add the sample buffer to the cell lysate and boil it at 95 °C for 10 min.
7. Load 10 μL of the samples onto 12% SDS-polyacrylamide gel plates, and conduct electrophoresis at 20 mA /gel (constant current).
8. Transfer the proteins in the gel onto a nitrocellulose membrane.
9. To block the non-specific binding of antibodies, incubate the membrane with the blocking buffer for 1 h at 4 °C.
10. Discard the old blocking buffer, and replace it with the blocking buffer containing the primary antibodies: anti-FLAG, His, V5, and β-actin.
11. Wash the membrane three times with TBST for 5 min each.
12. Gently shake the mixture for 1 h at 4 °C.
13. After four times washes with TBST at RT, shake the membrane for another 4 min.
14. Add TBST containing 1% skim milk and 2 μL of the secondary antibody linked to HRP: Anti-Mouse IgG, Anti-Rabbit IgG.
15. Gently shake the mixture for 1 h at 4 °C.
16. Wash the membrane four times with TBST at RT.
17. Incubate the membrane with the chemiluminescent detection reagent.
18. Detect the luminescence using an image analyzer.

3.7 Live Imaging for Evaluating the Reactivity of Exogenous Targets

1. Culture HEK293T cells in a glass-bottom dish.
2. Co-transfect the cells using expression vectors for the probe and the exogenous target RNA.
3. Twenty-four hours after transfection, change the culture medium to HBSS.

4. Add the luminescent substrate, and start the acquisition of luminescence and fluorescence sequentially using a luminescence microscope.
5. Analyze the luminescence and fluorescence intensity of the single cells.

3.8 Bioluminescence Microscope Observation of Living Cells

1. Culture NIH3T3 cells in 35 mM glass-bottom dishes.
2. Transfect the expression plasmids into the cells.
3. Incubate the cells for 24 h in DMEM containing 10% FBS at 37 °C in a 5% CO_2 atmosphere.
4. Exchange the medium to DMEM without FBS 24 h before the observation to induce serum starvation.
5. Change the medium to HBSS containing 2 ng/mL PDGF-BB.
6. Add the Nano-Glo Live Cell Substrate to the cell cultures to begin fluorescence and luminescence imaging.
7. Observe and analyze the cells using an inverted microscope.

3.9 Preparation of Primary Cultures of Rat Hippocampal Neurons

1. Dissect hippocampi from E21 rats (*see* **Note 10**) [24, 25].
2. Dissociate the cells by treating them with a preparation medium.
3. Plate the dissociated cells at a 6.0×10^5 cells/cm^2 density on a glass-bottom dish with plating medium.
4. Change the medium to the growth medium 3 h after the plating.
5. Maintain the cells with the feeding medium after 3 days (*see* **Note 11**).
6. Perform transfections using electroporation or lipofection (*see* **Note 12**).
7. Culture the transfected cells for 48 h.
8. Add the Nano-Glo Live Cell Substrate to the medium.
9. Before observation, change the cultivation medium to HBSS.
10. Add the Nano-Glo Live Cell Substrate into the medium,
11. Observe the cells using an inverted microscope.

4 Notes

1. NlucN-PUM1 was inserted between the *Nhe*I and *Not*I sites in the pcDNA4/V5-His ver. B vector. The same sequence was also introduced into the pColdI vector, specifically between the *Bam*HI and *Xba*I restriction sites.

2. mPUM2-NlucC was inserted between the *Nhe*I and *Eco*RI sites in the pcDNA4/V5-His ver. B vector. The same sequence was also introduced into the pColdI vector, specifically between the *Nde*I and *Sal*I restriction sites.
3. The 2A sequence contains self-cleaving sequences (amino acid sequence: GSGATNFSLLKQAGDVEENPGPELSSSSGS GEGRGSLLTCGDVEENPGP). The sequence was inserted between mPUM2-NlucC and NlucN-mPUM1.
4. Mix the Papain, DNase, and Hibernate-A solution immediately before use, and then incubate at 37 °C for 15 min.
5. Use an opaque microplate to measure bioluminescence intensities. When the high luminescence from the NLuc substrate hampers precise measurement, a black microplate, which minimizes signal leakage, is more suitable.
6. For uniform cell plating on a dish, resuspend the cells thoroughly by pipetting 5–10 times before seeding.
7. Use a multi-channel pipette for rapid cell seeding, preventing cell sedimentation that can lead to ununiform cell distribution across the wells.
8. To avoid a signal reduction, use a serum-free and phenol-red-free medium for bioluminescence measurements.
9. Perform all protein purification procedures at 4 °C to maintain protein stability and function.
10. Trim out the hippocampi to minimize the presence of non-neuronal cells, which could otherwise proliferate during the culturing process.
11. Transfection using lipofectamine or an electroporation system may result in notable cell toxicity or cell detachment from the dish. Ensure uniform distribution of the DNA mixture throughout the dish. If cell death is significant, optimize the amounts of transfected plasmid DNA and transfection methods.
12. Ensure to add a minimum of 1 mL of the growth medium to prevent the medium from drying out within a few days.

References

1. Moor AE, Golan M, Massasa EE, Lemze D, Weizman T, Shenhav R, Baydatch S, Mizrahi O, Winkler R, Golani O, Stern-Ginossar N, Itzkovitz S (2017) Global mRNA polarization regulates translation efficiency in the intestinal epithelium. Science 357(6357): 1299–1303
2. Das S, Vera M, Gandin V, Singer RH, Tutucci E (2021) Intracellular mRNA transport and localized translation. Nat Rev Mol Cell Biol 22(7):483–504
3. Martin KC, Ephrussi A (2009) mRNA localization: gene expression in the spatial dimension. Cell 136(4):719–730
4. Buxbaum AR, Haimovich G, Singer RH (2015) In the right place at the right time: visualizing and understanding mRNA localization. Nat Rev Mol Cell Biol 16(2):95–109

5. Rodriguez AJ, Shenoy SM, Singer RH, Condeelis J (2006) Visualization of mRNA translation in living cells. J Cell Biol 175(1): 67–76
6. Zhang HL, Singer RH, Bassell GJ (1999) Neurotrophin regulation of β-actin mRNA and protein localization within growth cones. J Cell Biol 147(1):59–70
7. Dasgupta S, Bader GD, Goyal S (2018) Single-cell RNA sequencing: a new window into cell scale dynamics. Biophys J 115(3):429–435
8. Yamada T, Yoshimura H, Inaguma A, Ozawa T (2011) Visualization of nonengineered single mRNAs in living cells using genetically encoded fluorescent probes. Anal Chem 83(14):5708–5714
9. Yoshimura H, Inaguma A, Yamada T, Ozawa T (2012) Fluorescent probes for imaging endogenous beta-actin mRNA in living cells using fluorescent protein-tagged pumilio. ACS Chem Biol 7(6):999–1005
10. Wang Y, Wang Z, Tanaka Hall TM (2013) Engineered proteins with Pumilio/fem-3 mRNA binding factor scaffold to manipulate RNA metabolism. FEBS J 280(16): 3755–3767
11. Thomas A (2015) Edwards, bespoke RNA recognition by Pumilios. Biochem Soc Trans 43(5):801–806
12. Yamada T, Yoshimura H, Shimada R, Hattori M, Eguchi M, Fujiwara TK, Kusumi A, Ozawa T (2016) Spatiotemporal analysis with a genetically encoded fluorescent RNA probe reveals TERRA function around telomeres. Sci Rep 6:38910
13. Yoshimura H (2018) Live cell imaging of endogenous RNAs using Pumilio homology domain mutants: principles and applications. Biochemistry 57(2):200–208
14. Wang X, McLachlan J, Zamore PD, Hall TMT (2002) Modular recognition of RNA by a human Pumilio-homology domain. Cell 110(4):501–512
15. Yoshimura H, Ozawa T (2016) Chapter Three – monitoring of RNA dynamics in living cells using PUM-HD and fluorescent protein reconstitution technique. In: Filonov GS, Jaffrey SR (eds) Methods in enzymology. Academic Press, pp 65–85
16. Yoshimura H, Ozawa T (2018) Real-time fluorescence imaging of single-molecule endogenous noncoding RNA in living cells. Methods Mol Biol 1649:337–347
17. Cheong CG, Hall TMT (2006) Engineering RNA sequence specificity of Pumilio repeats. Proc Natl Acad Sci 103(37):13635–13639
18. Ozawa T, Yoshimura H, Kim SB (2013) Advances in fluorescence and bioluminescence imaging. Anal Chem 85(2):590–609
19. Yoshimura H, Ozawa T (2014) Methods of split reporter reconstitution for the analysis of biomolecules. Chem Rec 14(3):492–501
20. Yang L, Nasu Y, Hattori M, Yoshimura H, Kanno A, Ozawa T (2013) Bioluminescent probes to analyze ligand-induced phosphatidylinositol 3,4,5-trisphosphate production with split luciferase complementation. Anal Chem 85(23):11352–11359
21. Eguchi M, Yoshimura H, Ueda Y, Ozawa T (2023) Split luciferase-fragment reconstitution for unveiling RNA localization and dynamics in live cells. ACS Sens 8(11):4055–4063
22. Dixon AS, Schwinn MK, Hall MP, Zimmerman K, Otto P, Lubben TH, Butler BL, Binkowski BF, Machleidt T, Kirkland TA, Wood MG, Eggers CT, Encell LP, Wood KV (2016) NanoLuc complementation reporter optimized for accurate measurement of protein interactions in cells. ACS Chem Biol 11(2): 400–408
23. Lu G, Dolgner SJ, Hall TM (2009) Understanding and engineering RNA sequence specificity of PUF proteins. Curr Opin Struct Biol 19(1):110–115
24. Bosch M, Castro J, Saneyoshi T, Matsuno H, Sur M, Hayashi Y (2014) Structural and molecular remodeling of dendritic spine substructures during long-term potentiation. Neuron 82(2):444–459
25. Matsuno H, Tsuchimine S, Fukuzato N, O'Hashi K, Kunugi H, Sohya K (2021) Sirtuin 6 is a regulator of dendrite morphogenesis in rat hippocampal neurons. Neurochem Int 145: 104959

Chapter 3

A Catalytic Hairpin Assembly-Based Approach for RNA Imaging in Living Cells

Aruni Karunanayake Mudiyanselage and Mingxu You

Abstract

The advancement of nucleic acid nanotechnology has resulted in broad applications of DNA- and RNA-based molecular sensors for bioanalysis. Catalytic hairpin assembly is such a type of programmable and enzyme-free nucleic acid circuit that has been popularly used in developing biosensors. Genetically encodable fluorogenic RNA-based devices have recently gained a lot of attentions as a powerful tool for intracellular imaging. Combining these two techniques together, we have developed a genetically encodable RNA-based catalytic hairpin assembly circuit for the highly sensitive detection of low-abundance RNAs in living cells. In this system, the binding of one target RNA can catalytically trigger the generation of tens-to-hundreds of fluorogenic RNA reporters and produce a significantly enhanced fluorescence signal. Here, we will introduce the detailed design, optimization, and imaging protocol of this new type of powerful imaging tools.

Key words Fluorescent sensor, Aptamer, Cellular imaging, Catalytic hairpin assembly, RNA circuit

1 Introduction

RNA-based fluorogenic sensors are genetically encodable and have gained its popularity in recent years for intracellular imaging [1]. These sensors are developed based on fluorogenic RNA aptamers and have been used to measure and quantify the cellular distribution, abundance, and flux of various metabolites, non-coding RNAs, and proteins [2–7]. Aptamers are short oligonucleotides that can fold and specifically recognize target molecules [8, 9]. The initially developed fluorogenic RNA aptamer, Spinach, also known as the RNA mimic of green fluorescence protein, is an RNA aptamer that specifically binds with a 5(*Z*)- 4H-imidazol-4-one, 5-[(3,5-difluoro-4-hydroxyphenyl)methylene]-3,5-dihydro-2,3-dimethyl (DFHBI) chromophore to produce green fluorescence [10]. After further improving the spectral and biophysical properties of Spinach, another widely used fluorogenic RNA

Shin-ichi Sato (ed.), *Live-Cell RNA Imaging: Methods and Protocols*, Methods in Molecular Biology, vol. 2875,
https://doi.org/10.1007/978-1-0716-4248-1_3,

aptamer, known as Broccoli, has been developed [11]. Broccoli exhibits improved thermal stability, and its cellular folding does not depend on high magnesium concentration.

Spinach and Broccoli have been used to develop allosteric and riboswitch aptamer sensors [12–14]. However, due to their moderate brightness and photostability, these first-generation RNA--based sensors do not have enough sensitivity toward the detection of low-abundance cellular targets [3, 15]. To improve the sensitivity of these fluorogenic RNA-based sensors, we have recently introduced the design of a "second-generation" RNA-based sensor termed as Catalytic Hairpin Assembly RNA circuit that is Genetically Encoded (CHARGE) [6]. In our design, Broccoli was divided into two non-fluorescent parts termed "Broc" and "Coli," which were, respectively, fused to the 3′ and 5′ ends of two RNA hairpins, H1 and H2 (Fig. 1a). Target binding induces the hybridization of H1 and H2, which further results in the reformation of Broccoli to turn on the fluorescence of DFHBI. Here, binding of one target can catalytically activate tens to hundreds of Broccoli fluorescence events.

This enzyme-free and isothermal signal amplification process [16] can provide robust fluorescence signal and enable sensitive imaging in living bacterial cells. CHARGE can also be potentially used as a platform for developing circuits for cellular regulation, diagnosis, and discovery of novel therapeutic agents, especially in cancer and infectious diseases [17–19]. In this chapter, we will discuss the detailed methods on the design, characterization, and optimization of CHARGE both in silico, in vitro, and inside living cells.

2 Materials

Prepare reagents and buffers using ultrapure water (prepared by Milli-Q system with a sensitivity of 18.2 MΩ cm at 25 °C) and molecular biology grade chemicals. Buffers and media need to be autoclaved or filter sterilized. All reagents and buffers are prepared at room temperature and stored at 4 °C (unless indicated otherwise). The Environment Health and Safety disposal regulations are followed when disposing waste materials.

2.1 In Vitro RNA Preparation and Characterization

1. 1 M Tris–HCl solution: Dissolving 60.57 g of Tris–Base in 500 mL of H_2O and adjust the pH to 8.0.
2. 500 mM EDTA solution: Dissolve 93.1 g of disodium EDTA•$2H_2O$ to 400 mL of H_2O. Stir and adjust the pH to 8.0 with NaOH (*see* **Note 1**). Make up to 500 mL with water and store at room temperature.

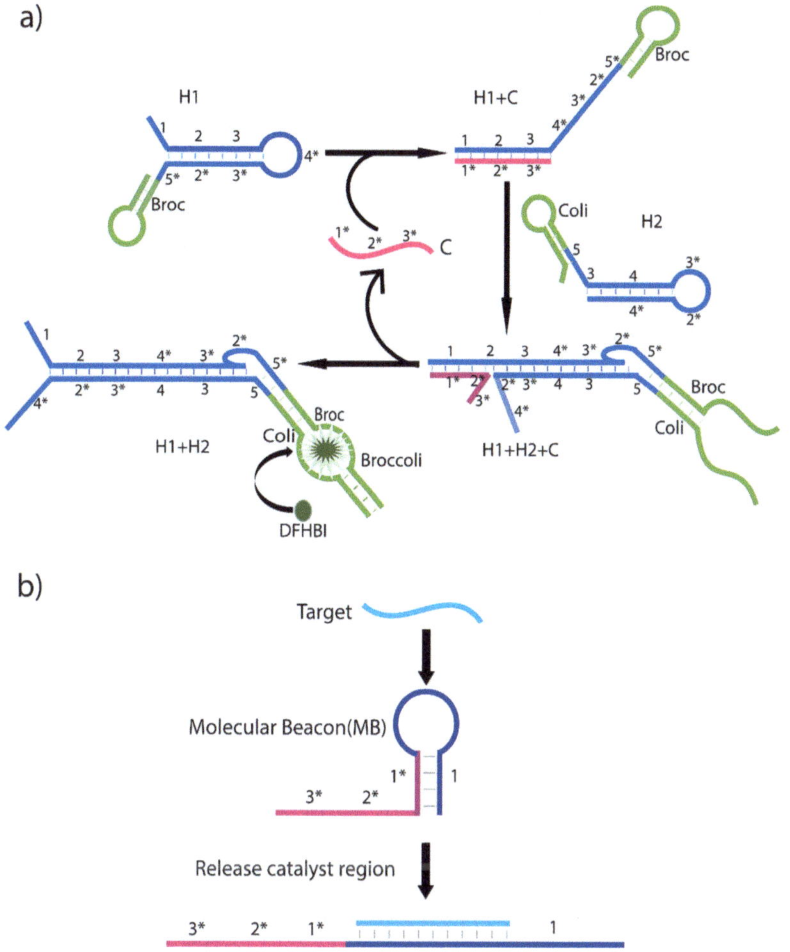

Fig. 1 Design principles of the CHARGE circuit. (**a**) The H1 and H2 strands are modified with split Broccoli, Broc and Coli, respectively. The catalyst (**c**) strand will initiate and catalyze the assembly of H1 and H2 into an H1 + H2 duplex. Recombined Broccoli in the H1 + H2 duplex activates the fluorescence of the DFHBI-1 T dye. (**b**) Modular CHARGE approach to detect various RNA targets. Binding of target RNA induces the opening of the molecular beacon, which further activates the catalyst sequence to initiate the assembly of H1 and H2, finally resulting in Broccoli fluorescence

3. 1× TE buffer: Mix 2 mL of above-made 0.5 M EDTA and 10 mL of Tris–HCl (pH 8.0), and add H_2O to 1000 mL.
4. 5× fluorescence assay and RNA assembly buffer: 50 mM Tris–HCl, 25 mM $MgCl_2$, 500 mM KCl, and 50 mM NaCl at pH 7.5.
5. 5× Broccoli folding buffer: 200 mM HEPES, 500 mM KCl, 0.5% DMSO, pH 7.4.
6. DFHBI-1 T stock solution: Dissolve DFHBI-1 T in 100% DMSO to make 20 mM stock solution. Further dilute into 2 mM DFHBI-1 T stock by 70–100% DMSO, and make 50–100 μL aliquots. Store at −20 °C (*see* **Note 2**).

7. Phusion high-fidelity DNA polymerase (2 U/μL). Store at −20 °C.
8. Monarch® PCR & DNA cleanup kit. Store at room temperature.
9. HiScribe™ T7 high yield RNA synthesis kit. Store at −20 °C.
10. DNase I (RNase-free) enzyme. Store at −20 °C.
11. TRIzole reagent. Store at room temperature (*see* **Note 3**).
12. RNase inhibitor, Murine. Store at −20 °C (*see* **Note 4**).
13. Agarose gel extraction and PCR purification kit.
14. Sephadex™ G-25 column to clean RNA: Mix Sephadex™ G-25 medium beads with 1× TE buffer (*see* **Note 5**). Use empty column (remove filter from commercial protein purification columns), and put silanized glass wool (*see* **Note 6**) at the bottom of the column. Pack the column by adding the beads and TE buffer mixture and centrifugation at 13,000 rpm at 25 °C (*see* **Note 7**). Wash four times with 1× TE buffer by centrifuging at 3750 rpm at 25 °C for 2 min in each wash. Remove residual buffer by centrifuging at 3750 rpm at 25 °C for 4 min before adding the sample for buffer exchange (*see* **Note 8**).

2.2 Materials for Molecular Cloning

1. LB medium: Dissolve 25 g of Difco™ LB broth powder in 1 L of ultrapure water. Autoclave at 121 °C for 30 min and store at room temperature.
2. Ampicillin stock solution: Dissolve ampicillin in autoclaved ultrapure water to make 150 mg/mL stock solution. Filter sterilize by passing through 0.22 μm syringe filter. Aliquot and store at −20 °C.
3. Streptomycin stock solutions: Prepare 75 mg/mL stock solution in autoclaved ultrapure water, and filter sterilize by passing through 0.22 μm syringe filter. Aliquot and store at −20 °C.
4. IPTG (Isopropyl β-D-1-thiogalactopyranoside) stock solutions: Prepare 1 M IPTG stock solution in autoclaved ultrapure water. Filter sterilize by passing through 0.2 μm syringe filter. Aliquot and store at −20 °C.
5. pETDuet and pCDFDuet vectors.
6. One-shot BL21 Star™ (DE3) chemically competent *E. coli*. Stored at −80 °C.
7. TOP10 chemically competent cell (*see* **Note 9**).
8. NdeI, PacI, EcoNI, and HindIII restriction enzymes. Store at −20 °C.
9. Antarctic phosphatase. Store at −20 °C.
10. QIAquick® gel extraction kit. Store at room temperature.

11. T4 DNA ligase. Store under −20 °C.
12. S.O.C. medium. Store at room temperature.
13. GeneJET plasmid miniprep kit. Store at room temperature.
14. Gibson Assembly® cloning kit.

2.3 Materials for Intracellular Imaging

1. Poly-L-lysine solution. Aliquot for single time use and store at −20 °C.
2. Dulbecco's phosphate-buffered saline (DPBS) buffer, no calcium, no magnesium.

3 Method

3.1 In Silico Design of CHARGE

NUPACK and mfold free online software can be used to design and optimize the sequence of Broc-H1 and Coli-H2 strands in silico. The folding and secondary structure of these strands are simulated both in the presence and absence of the target RNA sequence (*see* **Note 10**). Here are some general guidelines for the H1 and H2 hairpin design with split Broccoli fused into them.

1. Keep H1 stem at least 16 bases long and H1 loop 11–18 bases long. For both regions, GC% is in the range of 40–50%. Use NUPACK to simulate and check probability of H1 dimerization, which should be minimal (0%). Also make sure the H1 toehold region (domain "1" in Fig. 1, with 8–10 bases) is free from secondary structures (*see* **Note 11**).
2. Keep H2 stem 10–12 bases long and H1 loop 11–18 bases long. For both regions, GC% is in the range of 40–50%. Use NUPACK to simulate and check probability of H2 dimerization, which should be minimal (0%). Also make sure the H2 toehold region (domain "3, 5" in Fig. 1, with 8–10 bases) is free from secondary structures.
3. Split Broccoli aptamer is fused into H1 and H2 strands. Here, terminal stem loop sequence of the Broccoli aptamer can be completely removed to obtain the split version of the aptamer. The sequence of terminal stem region can be slightly modified to avoid unnecessary binding interactions. Here is one example of the "Broc" sequence (5′-*CAGAGAGACG-GUCGGGUCCACAGUUUCC* -3′) *and* the corresponding "Coli" sequence (5'-*GGAAACUGUGUCGAGUAGAGU-GUGGGCUCUCUG*-3′)
4. Add the Broc sequence into the 5′-end of the H1 strand and Coli sequence into the 3′-end of the H2 strand. Using NUPACK or mfold to ensure the fused Broc and Coli sequence, do not interfere with the predicted structures of H1 and H2.

5. Use NUPACK to simulate the hybridization efficiency and secondary structure of the H1 and catalyst (C) complex. Ensure a high percentage of duplex formation (>90%) and the H2 toehold binding region (domain "3*2*5*" in Fig. 1) is free from secondary structures.
6. To further develop a modular CHARGE system to detect various RNA targets, a molecular beacon structure should be further in silico designed where it contains the target RNA interacting region (12–18 bases) in the loop region and the catalyst RNA sequence embedded in the stem region (6–8 base pairs). Use NUPACK to ensure the binding of target RNA will induce the opening of molecular beacon and free the catalyst region to initiate the assembly of H1 and H2.
7. Once the hairpin sequences and target/catalyst sequences are selected, a T7 promoter sequence (5′-*TAATACGACTCACTATAG*-3′) should be added to the 5′-end of the DNA non-template strand for successful in vitro transcription. Meanwhile, if these strands will be further cloned into a vector system via restriction enzyme digestion, appropriate restriction enzyme recognition sequences should be added to the 3′- and 5′-ends of the insert. Design appropriate forward and reverse primers for PCR or a non-template strand for direct transcription. These DNA oligonucleotides can be commercially synthesized and purified.

3.2 In Vitro Synthesis and Characterization of CHARGE

1. Dissolve commercially synthesized oligonucleotides in 1× TE buffer (10 mM Tris-HCl, 0.1 mM EDTA buffer at pH 7.5) to obtain the final concentration of 100 μM, and store at −20 °C.
2. Prepare double-stranded DNA templates for in vitro transcription using standard PCR protocol. Follow manufacturer's instruction, mix appropriate amount of Phusion high-fidelity DNA polymerase, DNA templates, forward and reverse primes, and other required reagents in the PCR tubes, and amplify in a thermocycler.
3. Verify the product in 2% agarose gel, and purify using a QIAquick PCR purification kit. If the PCR products are not sufficiently pure, a gel purification is needed. Measure the concentration of the purified PCR product with a NanoDrop.
4. Follow manufacturer's protocol to transcribe RNAs in vitro using a HiScribe™ T7 high-yield RNA synthesis kit. Mix appropriate amount of T7 RNA polymerase, and purify DNA with other required reagents at 20–40 μL scale in a microcentrifuge tube. Incubate under 37 °C for 4 h in a thermomixer.
5. After the transcription, add appropriate amount (2 units per 10 μg of RNA) of DNase I and buffer directly to the above T7 transcription mixture, and incubate under 37 °C for 30 min in

the thermomixer. Digestion with DNase I will remove template DNA strands from the transcription mixture. Denature the enzymes at 72 °C for 10 min using a dry bath.

6. Purify the in vitro transcribed RNAs using a home-made Sephadex™ G-25 column (*see* **Note 12**). Validate the size of RNAs, and gel purify the desired RNA product with 10% urea-denatured polyacrylamide gel electrophoresis (*see* **Note 13**). Add Murine RNase inhibitor according to the manufacturer's instruction. Measure the concentration of obtained RNAs using NanoDrop. All the RNA molecules are then dispensed into aliquots with RNase-free water or TE buffer and stored at −20 °C for immediate usage or at −80 °C for long-term storage.
7. To ensure the proper folding of each RNA strand, before each test, RNA molecules are pre-heated at 95 °C for 3 min and then slowly cooled to 25 °C at the rate of −3 °C/min.
8. To test the in vitro performance of CHARGE, conduct the fluorescence assay in an RNA assembly buffer (10 mM Tris–HCl, 5 mM $MgCl_2$, 100 mM KCl, and 10 mM NaCl at pH 7.5) (*see* **Note 14**). In these measurements, 100–500 nM H1 and H2 can be used with the addition of 0.1–250 nM target RNA. All these reactions are initiated by adding either target or H1 to the mixture of other RNA strands and dyes (5 μM DFHBI-IT). The obtained fluorescence results should be compared to the fluorescence of Broccoli (positive control) measured at the same concentration level under the same experimental conditions (*see* **Note 15**) (Fig. 2a).
9. For the dose-response study, well-mixed RNA samples are incubated for ~2 h before taking the fluorescence measurements (Fig. 2b). Similarly, the kinetics can be tested by monitoring the signal increase over 2 h after adding the target RNA (*see* **Note 16**). The optimal sensor should exhibit a large increase in fluorescence and faster kinetics (*see* **Note 17**).

3.3 Cloning of the CHARGE Sensors into E. coli *Cells*

1. H1 and H2 hairpins can be cloned into a dual vector such as pETDuet in a sequential manner (*see* **Note 18**). Standard restriction site cloning protocol or Gibson assembly protocol can be followed here. We will use a standard double restriction digestion protocol as an example here to illustrate how H2 can be cloned into the pETDuet vector with *Nde* I and *Pac* I restriction enzymes (*see* **Note 19**).
2. Add 10 units of Antarctic phosphatase and 1× phosphatase reaction buffer (*see* **Note 20**) into the digested pETDuet plasmid. Incubate at 37 °C for 30 min and heat inactivation at 80 °C for 2 min. Run the sample in 1% agarose gel (*see* **Note 21**), and purify the intended product.

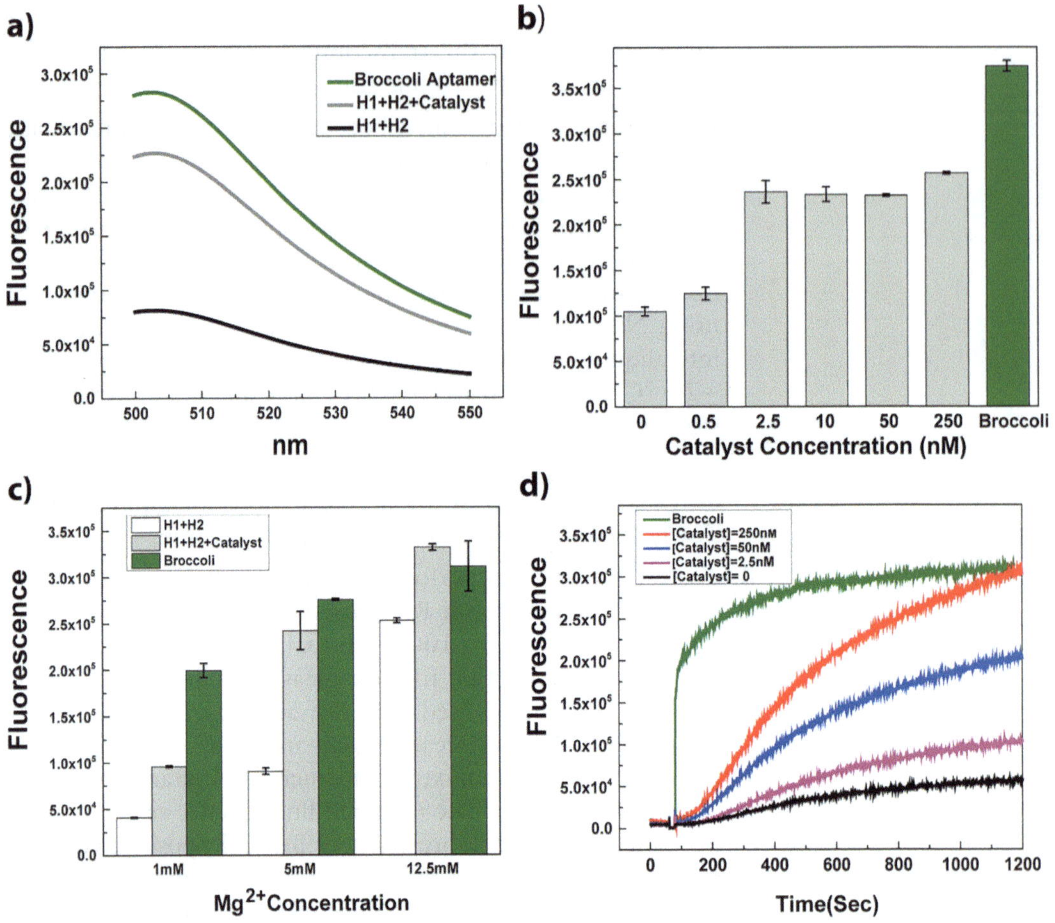

Fig. 2 Sample results for the in vitro fluorescence measurement of CHARGE. (**a**) Fluorescence emission spectra of Broccoli aptamer, CHARGE circuit in the presence or absence of CHARGE. (**b**) Fluorescence response of CHARGE in response to different concentrations of catalyst

3. PCR amplify H2 with primers containing *NdeI* or *PacI* restriction sites at the 5′- and 3′-end, respectively. Follow the **steps 1–3** as described in Subheading 3.2 for PCR amplification and purification. Conduct the restriction digestion (**step 1** in Subheading 3.3), and purify the product using the PCR cleanup kit (*see* **Note 22**).
4. Ligate the digested pETduet plasmid with the H2 insert at a molar ratio of 1:3 using T4 DNA ligase enzyme (*see* **Note 23**). Prepare a negative ligation mixture without adding any insert. Incubate samples at 16 °C overnight in a thermomixer with 300 rpm shaking. Inactivate the enzymes at 65 °C for 10 min.
5. The ligated products are then transformed into *E. coli* cells such as TOP10. Thaw two tubes of 50 μL of TOP10 chemical

competent cells on ice. Carefully transfer cells into the bottom of polypropylene BD round bottom 15 mL tubes. Add 4 μL of ligated product into each tube, mix by gently shaking on ice, and incubate on ice for 20–30 min (*see* **Note 24**).

6. After the incubation, conduct the transformation by a heat shock method. Here incubate the tubes containing cells and DNA in a 42 °C water bath for 1 min (*see* **Note 25**). Afterward, immediately place the tubes back in the ice for 2 min. Add 300–500 μL S.O.C. medium, and incubate at 37 °C for 2 h with shaking at 200 rpm.
7. Plate these cells on LB agar/Amp (150 μg/mL ampicillin) petri plates, and incubate at 37 °C for 12–14 h until individual colonies can be easily identified.
8. Randomly pick 10–12 colonies, and grow respectively in 5 mL LB/Amp liquid media at 37 °C for 12–14 h with shaking at 200 rpm.
9. Extract the plasmid from each inoculation solution using a GeneJET plasmid extraction kit. Perform Sanger sequencing to confirm the correct insertion in the plasmid. Once the H2 insertion has confirmed, proceed to the H1 insertion step.
10. For the H1 insertion, traditional restriction enzyme digestion procedure can be performed if the double digestion compatible restriction sites are available. Here we use Gibson assembly protocol to clone H1 into pETduet vector because of difficulty in achieving compatible double digestion sites. H1 is cloned using a Gibson assembly® cloning kit. pETDuet vector was PCR amplified with primers designed to remove the target insertion region. Resulting product is linearized pETDuet vector and confirms the sequence of amplified PCR product by Sanger sequence. H1 insert sequence is designed such that it will amplify by 2–3 fragment overlap PCR during the Gibson assembly protocol. Linear vector and H1 insert with 15–20 bp overlap end mix according to the Gibson assembly kit guidelines. Incubate the samples in a thermocycler at 50 °C for 30 min (*see* **Note 26**), and store at −20 °C until ready for transformation.
11. Follow **steps 5–9** of Subheading 3.3 to confirm the H1 insertion into pETDuet vector.
12. As a control, the pETDuet vector containing only H1 is also prepared (*see* **Note 27**).
13. After the correct sequences of pETDuet-H1, pETDuet-H2, and pETDuet-H1/H2 have been confirmed, these plasmids are transformed into *E. coli* imaging cell lines such as BL21 Star™ (DE3) (New England Biolabs) (*see* **Note 28**). Thaw

three tubes of chemically competent BL21 Star™ (DE3) cells in ice. Then add ~40 ng of pETDuet-H1, pETDuet-H2, and pETDuet-H1/H2, respectively, into one tube. Transform as described in **steps 6–7** (Subheading 3.3), and screen colonies against ampicillin resistance. Randomly select 10–12 colonies, and confirm the correct transformation by Sanger sequencing.

14. The catalyst and molecular beacon strand can be cloned into another pETDuet compatible dual-expressing vector, such as pCDFDuet. Digest the PCR product and vector with *Eco*N I and *Hind* III restriction enzymes, and follow **steps 2–6** (Subheading 3.3) to produce ligation product. Grow transformed cells in LB agar plate with streptomycin (75 μg/mL), and screen colonies for streptomycin resistance. Randomly isolate 10–12 colonies, purify plasmids using GeneJET plasmid purification kit, and confirm the insertion of catalyst and molecular beacon by Sanger sequencing.
15. Heat shock to transform the pCDFDuet-catalyst or the pCDFDuet-molecular beacon plasmid into the above BL21 Star™ (DE3) cells that express pETDuet-H1, pETDuet-H2, or pETDuet-H1/H2 vectors. Screen colonies with both ampicillin and streptomycin resistance. Confirm the correct transformation by Sanger sequencing.
16. As a positive control, prepare BL21 Star™ (DE3) cells expressing the pETDuet-Broccoli vector following the above protocol, and select the colonies against ampicillin resistance. Confirm the correct transformation by Sanger sequencing.
17. Prepare glycerol stocks of each BL21 Star™ (DE3) cell line as generated above, and store at −80 °C for long-term storage (*see* **Note 29**).

3.4 Fluorescence Imaging of CHARGE Sensors in Live *E. coli* Cells

1. Take the freezer stocks of BL21 Star™ (DE3) cells, and prepare streak agar plates to generate single colonies of cells expressing pETDuet-H1, pETDuet-H2, pETDuet-H1 + pCDFDuet-catalyst, pCDFDuet-catalyst, pETDuet-H2 + pCDFDuet-catalyst, pETDuet-H1/H2 + pCDFDuet-catalyst, and pETDuet-H1/H2 + pCDFDuet-molecular beacon, respectively. 150 μg/mL ampicillin is added in the agar plates for growing cells expressing pETDuet-H1, pETDuet-H2, or pETDuet-H1/H2. 75 μg/mL streptomycin is added in the plates for growing cells expressing pCDFDuet-catalyst or pCDFDuet-molecular beacon.
2. Pick a single colony, and grow in 5 mL LB medium with appropriate antibiotic for 12 h overnight at 37 °C with shaking at 200 rpm.

3. Take a 1:10 dilution of the overnight cell culture, and measure its optical density at 600 nm (OD_{600}). Here, the final OD_{600} should be less than 2.
4. Inoculate 50 μL of the overnight grown culture into a 2 mL LB medium with appropriate antibiotic, and grow at 37 °C at 200 rpm until OD_{600} reaches 0.5–0.6.
5. Induce RNA expression with 1 mM isopropyl β-D-1-thiogalactopyranoside (IPTG), and continue to grow for additional 2 h. If the cells are grown for the *SgrS* detection (*see* **Note 30**), cells can be IPTG induced for 1.5 h, and then treat with 1 g/L or 20 g/L glucose for an additional 30 min.
6. While growing the cells, add 300 μL poly-L-lysine (PLL) solution (0.1% w/v) into each well of a glass-bottomed 8-well chamber imaging plate (Cellvis, CA). Incubate the plate at 37 °C for 3 h, remove the PLL solution, and rinse each well twice with 500 μL sterile water to remove excess PLL. Add 500 μL sterile water, and store the plate at room temperature before usage (*see* **Note 31**).
7. Take 200 μL of an IPTG-induced culture (**step 6**), and spin at 5000 g for 2 min at room temperature to pellet the culture. Remove the media and resuspend the cell pellet in 1 mL of 1× DPBS.
8. Transfer 200 μL of the resuspended culture into the PLL-coated plate, and incubate at 37 °C for 45 min to allow the cells to adhere onto the surface of the imaging plate (*see* **Note 32**). After the incubation, carefully wash out unattached cells by washing each well twice with 500 μL of 1× DPBS. Add 200 μL of 1× DPBS supplemented with 200 μM DFHBI-1 T into each well of the imaging plate. Incubate the plate at room temperature for 1 h before imaging.
9. Cell imaging can be performed by exciting with a 488 nM laser through a 60× oil immersion objective. The following filter sets can be used as an example: 470/40 nM excitation filter, a 496 nM dichroic mirror (long pass), and a 525/50 nM emission filter. Here, a Nikon Eclipse Ti microscope can be used. Usually, the proper exposure time is set in the range of 100–2000 ms, which allows a bright cellular fluorescence signal without saturated pixels (*see* **Note 33**). Always start the imaging with cells expressing Broccoli aptamer, i.e., the positive control (Fig. 3).
10. Analyze the confocal imaging data using softwares like NiS-Elements AR and ImageJ. Quantitative data calculation and fitting can be done using software like Origin (*see* **Note 34**).

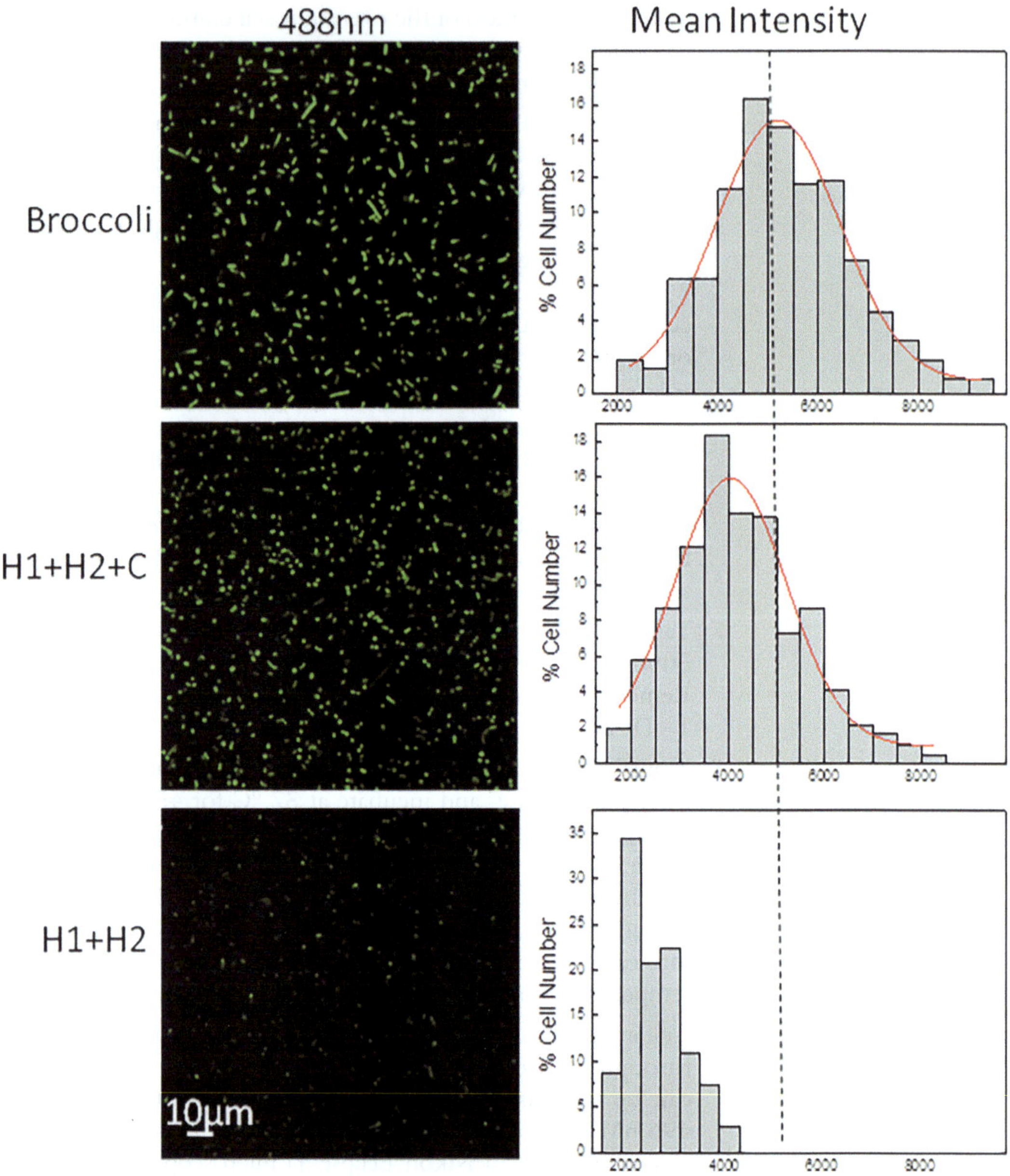

Fig. 3 Example of confocal cellular imaging of CHARGE with or without the catalyst and cells expressing Broccoli are used as a positive control. Distributions of cellular fluorescence levels are analyzed and shown on the right

4 Notes

1. Disodium EDTA•$2H_2O$ will not be fully dissolved until the pH of the solution is ~8 by adding NaOH.
2. DFHBI-1 T can be dissolved in 100% DMSO to prepare 20 mM stock. High concentration of DMSO may be toxic to

the cells, and therefore we suggest having the final working concentration of DMSO to be at most 1% or 100-fold diluted from the stock solution.

3. Trizole reagent is required only if cellular RNA is going to be extracted.
4. Add RNase inhibitor after the transcription and gel purification, and for long-term storage at −20 °C or −80 °C, RNase inhibitor helps to prevent RNA from degradation.
5. Soak the Sephadex™ G-25 medium beads in 1× TE buffer, and store at 4 °C overnight. Mix the beads with the TE buffer every time, and leave it under 4 °C overnight before usage. Swirl the mixture right before usage to resuspend the beads in the TE buffer.
6. Remove the cellulose filter from the column because RNA can be adsorbed onto these cellulose filters. Put glass wool at the bottom of the column to prevent the Sephadex beads from getting leak into the column. To prevent RNA from getting adsorbed or adhering onto the glass wool, silanize the glass wool before usage by immersing in Sigmacote™ (Sigma Diagnostics, Livonia, MI, USA) in a chemical hood, and wait until the glass wool is completely dry.
7. To pack the column, add TE buffer-soaked beads into the column, centrifuge, and discard the supernatant. Repeat until the 80–90% of column volume gets packed with beads.
8. Make sure the packed column is completely dry before transferring the RNA mixture into the column. This lab-made column will effectively clean RNAs with sizes smaller than 100 bp.
9. Competent cells can be self-prepared or purchased from companies such as New England Biolabs.
10. These structure prediction programs can help identify the correctly folded H1 and H2 strands and optimize the sequences of H1 and H2 to avoid nonspecific strand binding.
11. Structure prediction program such as NUPACK can provide the probability of obtaining the intended structures and help identify how well the design of hairpins can assemble into the desired structures.
12. The RNA purification can be performed with commercially available kits as well (Monarch RNA Cleanup Kit, New England Biolabs). The cut-off molecular weight of the column in the kits should be smaller than that of the RNA product.
13. A previously reported protocol [20] can be followed to purify RNA from denature PAGE gels, and then the RNAs can be

isolated following standard ethanol precipitation protocols [21].

14. Efficient folding of Broccoli requires 1–5 mM of magnesium ions, which is similar to the intracellular levels.
15. Perform the fluorescence measurement of both positive and negative controls at the same time with the CHRAGE sensor. The negative control only contains DFHBI-1 T in 1× RNA folding buffer and $MgCl_2$, without adding any targets or RNAs. In the positive control, Broccoli RNA is used instead of the sensor strands. In addition, the background fluorescence and signal leakage can be determined by using solutions containing H1 + H2, H1 + catalyst, and H2 + catalyst.
16. CHARGE reaction requires ~2 h to reach maximum fluorescence signal. In general, a 500–600 nM range of emission spectra is collected with an excitation at 480 nM. The kinetic assays can be conducted using 480 nM excitation and 503 nM emission.
17. If the designed sensor does not exhibit noticeable target-induced fluorescence enhancement, the stem and loop sequence of the H1 and H2 strands should be optimized (by changing sequence of stem and loop region to achieve the criteria mentioned in **steps 1–2** of Subheading 3.1) to achieve a sensor that can produce at least two–threefold of fluorescence enhancement.
18. When choosing the restriction enzymes, make sure the enzyme recognition sites only exist at the desired location and do not disrupt the transcription reading frame after the insertion. Simultaneously digesting the DNA strands with two restriction enzymes is time-saving and preferred. Tools like NEB Double Digest Finder can suggest optimized parameters for the double digestion, including buffer conditions, inactivation temperature, methylation sensitivity, etc.
19. Follow the product specifications and NEB Double Digest Finder protocol to achieve optimal digestion result. Here, 1 μg of DNA is mixed with 10 units of both NdeI and PacI enzymes in NEB buffer 1.1 at 50 μL scale, and the mixture is incubated at 37 °C for 3 h. Inactivate the enzyme by incubating at 65 °C for 5 min.
20. Treat with phosphatase to remove the 5′ phosphate group from the digested vector product, and prevent the vector from self-ligation during the ligation step.
21. Run an agarose gel to confirm the successful restriction digestion. Double digestion will result in two product bands. To resolve the products properly, the percentage of agarose gel can be altered depending on the size of DNA.

22. For the efficient digestion of short PCR DNA insert (<200 bases), use a smaller amount of DNA (e.g., 300 ng).
23. Follow the product specification and guideline of the T4 DNA ligase to set up the ligation reaction. The molar ratio of vector to insert varies with the size of the DNA insert/vector size. An NEBiocalculator tool can be used to determine the molar ratio. For example, if the insert is shorter than 200 bases, a molar ratio of 1:5 can increase the insertion efficiency as compared to that of 1:3.
24. Do not mix by pipetting because air bubbles can hinder the transformation efficiency.
25. Do not mix while the tubes are still in the water bath.
26. The number of Gibson assembly fragments varies with the chosen overlap PCR sites, which is depending on the total length of the DNA strand. If 4–6 fragments are assembled, incubate for 60 min.
27. In the case that the H1 insertion is difficult, first prepare a pETDuet vector with an H1 insert using the Gibson assembly protocol. Then follow traditional restriction enzyme digestion procedure to clone the H2 insert.
28. BL21 Star™ (DE3) cells carry a mutated *rne* gene that encodes a truncated RNase E enzyme. It lacks the ability for RNA degradation, resulting in high RNA sensor concentrations, which can potentially exhibit high fluorescence intensity.
29. To prepare the glycerol stocks, grow bacterial colonies in LB media until OD_{600} reaches 0.5–0.6. Mix the grown culture with autoclaved 100% glycerol to reach the final glycerol percentage of 20%, and flash freeze with liquid nitrogen.
30. *SgrS* is an Hfq protein-dependent small RNA that is expressed in response to glucose stress. We have previously designed a modular CHARGE sensor to sense *SgrS* RNA in bacterial cells [6].
31. The same procedure can be applied to a 24-well plate as well. These treated wells can be stored with sterile water for 24 h at room temperature or 1 week at 4 °C. Make sure to cover the plate with parafilm to prevent it from drying out.
32. Here, set up the negative and positive controls, and repeat wells in the same imaging plate.
33. These imaging parameters can be adjusted and optimized based on the confocal microscope used for the imaging.
34. Here, different data analysis program can be used (e.g., MATLAB).

Acknowledgments

The authors gratefully acknowledge the UMass Amherst start-up grant, NSF CAREER, Sloan Research Fellowship, and Camille Dreyfus Teacher-Scholar Award to M. You.

References

1. Karunanayake Mudiyanselage APKK, Wu R, Leon-Duque MA et al (2019) "Second-generation" fluorogenic RNA-based sensors. Methods 161:24–34
2. Bouhedda F, Autour A, Ryckelynck M (2018) Light-up RNA aptamers and their cognate fluorogens: from their development to their applications. Int J Mol Sci 19:44
3. Ouellet J (2016) RNA fluorescence with light-up aptamers. Front Chem 4:1–12
4. Wu R, Karunanayake Mudiyanselage APKK, Ren K et al (2020) Ratiometric fluorogenic RNA-based sensors for imaging live-cell dynamics of small molecules. ACS Appl Bio Mater 3:2633–2642
5. Ren K, Wu R, Karunanayake Mudiyanselage APKK et al (2020) In situ genetically cascaded amplification for imaging RNA subcellular locations. J Am Chem Soc 142:2968–2974
6. Karunanayake Mudiyanselage APKK, Yu Q, Leon-Duque MA et al (2018) Genetically encoded catalytic hairpin assembly for sensitive RNA imaging in live cells. J Am Chem Soc 140: 8739–8745
7. Yu Q, Shi J, Mudiyanselage APKKK et al (2019) Genetically encoded RNA-based sensors for intracellular imaging of silver ions. Chem Commun 55:707–710
8. Holeman LA, Robinson SL, Szostak JW et al (1998) Isolation and characterization of fluorophore-binding RNA aptamers. Fold Des 3:423–431
9. Song KM, Lee S, Ban C (2012) Aptamers and their biological applications. Sensors 12:612–631
10. Paige JS, Wu KY, Jaffrey SR (2011) RNA mimics of green fluorescent protein. Science 333:642–646
11. Filonov GS, Moon JD, Svensen N et al (2014) Broccoli: rapid selection of an RNA mimic of green fluorescent protein by fluorescence-based selection and directed evolution. J Am Chem Soc 136:16299–16308
12. Strack RL, Song W, Jaffrey SR (2014) Using spinach-based sensors for fluorescence imaging of intracellular metabolites and proteins in living bacteria. Nat Protoc 9:146–155
13. You M, Litke JL, Jaffrey SR (2015) Imaging metabolite dynamics in living cells using a spinach-based riboswitch. Proc Natl Acad Sci 112:E2756–E2765
14. Wu R, Karunanayake Mudiyanselage APKK, Shafiei F et al (2019) Genetically encoded ratiometric RNA-based sensors for quantitative imaging of small molecules in living cells. Angew Chem Int Ed 58:18271–18275
15. Strack RL, Jaffrey SR (2013) New approaches for sensing metabolites and proteins in live cells using RNA. Curr Opin Chem Biol 17:651–655
16. Jung C, Ellington AD (2014) Diagnostic applications of nucleic acid circuits. Acc Chem Res 47:1825–1835
17. Liu J, Zhang Y, Xie H et al (2019) Applications of catalytic hairpin assembly reaction in biosensing. Small 15:1–17
18. Molefe PF, Masamba P, Oyinloye BE et al (2018) Molecular application of aptamers in the diagnosis and treatment of cancer and communicable diseases. Pharmaceuticals 11:1–24
19. Fu P, Xing S, Xu M et al (2020) Peptide nucleic acid-based electrochemical biosensor for simultaneous detection of multiple microRNAs from cancer cells with catalytic hairpin assembly amplification. Sensors Actuators B Chem 305: 127545
20. Piganeau N (2009) In vitro selection of allosteric ribozymes. Methods Mol Biol 535:45–57
21. Green MR, Sambrook J (2020) Precipitation of RNA with ethanol. Cold Spring Harb Protoc 2020:89–91

Chapter 4

Live Single-Cell Transcriptional Dynamics in Plant Cells

Sahar Hani, Caroline Mercier, Pascale David, Edouard Bertrand, Thierry Desnos, and Laurent Nussaume

Abstract

Transcriptional reprogramming plays a key role in a variety of biological processes. Recent advances in RNA imaging techniques have allowed to visualize, in vivo, transcription-related mechanisms in different organisms. The MS2 system constitutes a robust method that has been used for over two decades to image multiple steps of a transcript's life cycle from "birth to death" with high spatiotemporal resolution in the animal field. It is based on the high affinity binding of the MS2 bacteriophage coat protein to its RNA hairpin ligands. Despite its broad applicability, a limited number of studies have implemented the system in plants, but without exploiting its full potential. Here, we describe the transposition of the MS2 technique to *Arabidopsis*. Combined with microfluidics, it allows to visualize the transcriptional repression of a phosphate starvation induced gene (*SPX1*) upon phosphate refeeding in vivo. The system provided access to the transcriptional response kinetics of individual cells, gene expression heterogeneity, and revealed bursting phenomena in plantae. The described methods provide new insights for multiple applications.

Key words MS2, MCP, RNA, In vivo, Transcription, Imaging, Microfluidic, Plant, Phosphate

1 Introduction

As sessile organisms, plants start their lives as heterotrophs, and then light promotes massive changes allowing them to develop autotrophic structures to sustain their development. Throughout their lifetimes, they also experience continuous fluctuations arising from pathogens, abiotic factors, and an ever-changing environment. Thus, a plant's survival depends on its ability to accommodate to its surrounding through substantial internal reprogramming at various levels.

Transcriptional reprogramming plays a significant role in these coping mechanisms, as highlighted by the elevated number (5.4%) of coding sequences encoding transcription factors (TF)

Sahar Hani and Caroline Mercier have equally contributed to this chapter.

Shin-ichi Sato (ed.), *Live-Cell RNA Imaging: Methods and Protocols*, Methods in Molecular Biology, vol. 2875, https://doi.org/10.1007/978-1-0716-4248-1_4,

(https://agris-knowledgebase.org/) in *Arabidopsis thaliana* for instance. Similarly, genome sequencing underlined the importance of TFs in other plant species. For example, a total of 2448 has been reported in sorghum, along with 1611 in rice and 3337 in maize, representing 5%, 3%, and 4.5% of their coding sequences, respectively [1]. These data are quite similar to those found in animals; indeed for humans 6% of the coding sequences encode transcription factors [2].

Over the past decades, various techniques have been developed to evaluate transcriptional changes, most frequently in entire organs. However, by measuring the expression of RNAs in mixed cell populations, critical differences between individual cells may be obscured. Methods to access specific tissues or cells have been developed since. But they required specific cell marker expression and cell sorting or microdissection that are technically challenging and hinder fast kinetic studies.

Other approaches such as reporter gene assays allowed precise gene expression tracking coupled with cellular resolution [3]. Transcriptional fusions with fluorescent proteins such as GFP have been used to monitor specific gene's expression profile at the single-cell level. Nevertheless, fluorophores require time to mature and accumulate in order to be visualized, preventing (or delaying) access to fast processes. All of these features must be taken into account when studying rapid transcriptional modifications that occur within seconds [4–6].

Additionally, although these approaches provide valuable information, they quantify bulk mRNA abundance resulting from an equilibrium defined by the rate of transcript synthesis and its degradation. In this context, many fluorophores decay within hours, and RNA degradation in plants is highly variable between genes as an mRNA's half-life ranges from minutes to days. These characteristics further limit these techniques for tracking transcription kinetics and particularly transcriptional repression.

To circumvent these issues, considerable efforts were spent to establish methods for live RNA visualization. In this framework, the MS2-MCP system constitutes the prototypical RNA imaging technique. It is based on the high affinity binding of the viral bacteriophage MS2 coat protein (MCP) to its RNA hairpin ligand binding sites.

One of its first applications was for visualization of mRNA localization in living yeast [7]. For in vivo RNA labeling, a tandem repetition of a 19-nucleotide-long stem-loop sequence (MS2) was transcriptionally fused to the RNA under investigation. Detection is performed by co-expressing the fluorescently tagged MCP which recognizes and binds the MS2 loops (Fig. 1).

The high specificity and sensitivity of the technique allowed its applicability to expand rapidly. It became a standard method for studying many processes involving RNAs in microorganisms, yeast,

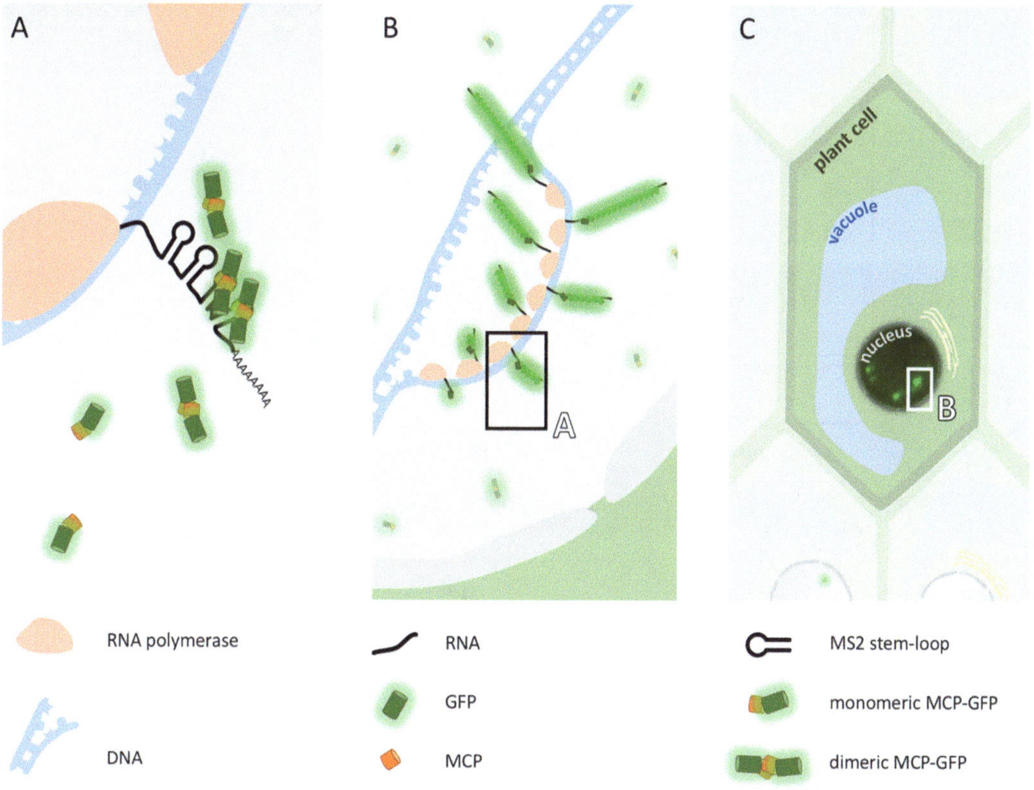

Fig. 1 Principle of the MS2-MCP system. (**a**) The MS2 RNA stem-loop motifs are included in the mRNAs undergoing transcription by the RNA Polymerase II. The MCP, translationally fused to the GFP, self-dimerizes, and then, one couple binds to one MS2 stem-loop. Each MS2 stem-loop can therefore carry two GFP. (**b**) Several mRNAs can be synthesized at the same time from one transcriptionally active gene. This results in the concentrated accumulation of a lot of GFP bound to MS2. These locally accumulated GFP make a fluorescent green spot visible in nuclei by microscopy (**c**)

and mammalian cells in culture, at the single-cell level with high spatial and temporal resolution. It enabled the dissection and analysis in vivo of all transcript's life steps from birth to death such as transcription, pre-mRNA processing and splicing [8], mRNA transport, localization [7], and degradation [9]. This technique has also been employed to investigate with details transcriptional dynamics, gene response to different signals (Fig. 2), bursting (Fig. 3), and transcriptional memory, thus highlighting the uniqueness of individual cells [10–14].

Despite its broad applicability, a limited number of studies have implemented the MS2 technology in multicellular organisms and in some rare attempts in plants but without exploiting its full potential. This is mainly due to the technical difficulties of transposing it into plantae and the insufficient number of MS2 repeats to produce sufficient signal. In addition to the cloning complexities, the optical properties of plant cells and tissues present considerable challenges

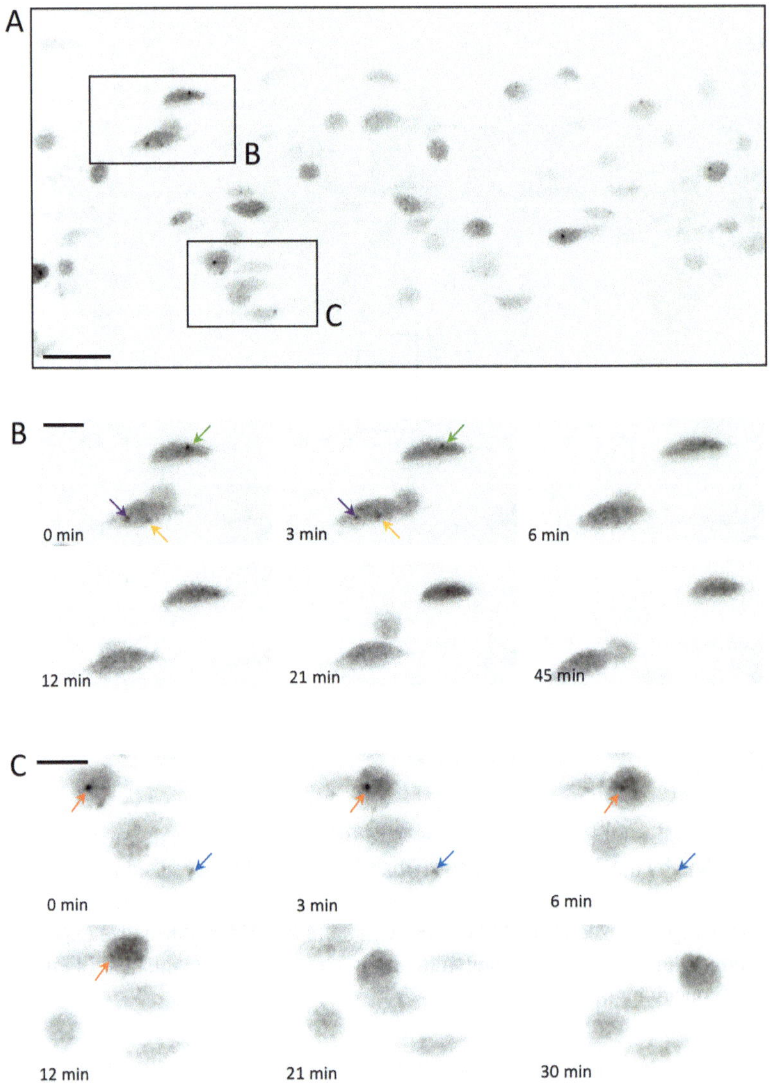

Fig. 2 Combining microfluidic and MS2 technology reveals fast transcriptional repression triggered by Pi supply in p*SPX1*::*MS2*x128 transgenic plants. (**a**) Maximum intensity projection from a time-lapse video recorded in 3D (200 z planes), which displays MCP-GFP fluorescence in the transgenic plants at the start of phosphate refeeding ($t = 0$). Scale bar, 20 μm. (**b, c**) Images are maximum intensity projections from the time-lapse video shown in (**a**), taken at the indicated time points after Pi refeeding and displaying MCP-GFP fluorescence. Scale bar, 10 μm

for the technique, as many molecules emit high levels of background and autofluoresce which adversely affect detection efficiency. Thus, its aims were restricted to detection of plant viruses [15], study of mRNA transport [16], and localization [17–19].

More recently, advances in the cloning strategies and imaging methods facilitated the application of the technique in living plants and allowed quantitative imaging of RNA polymerase II activity

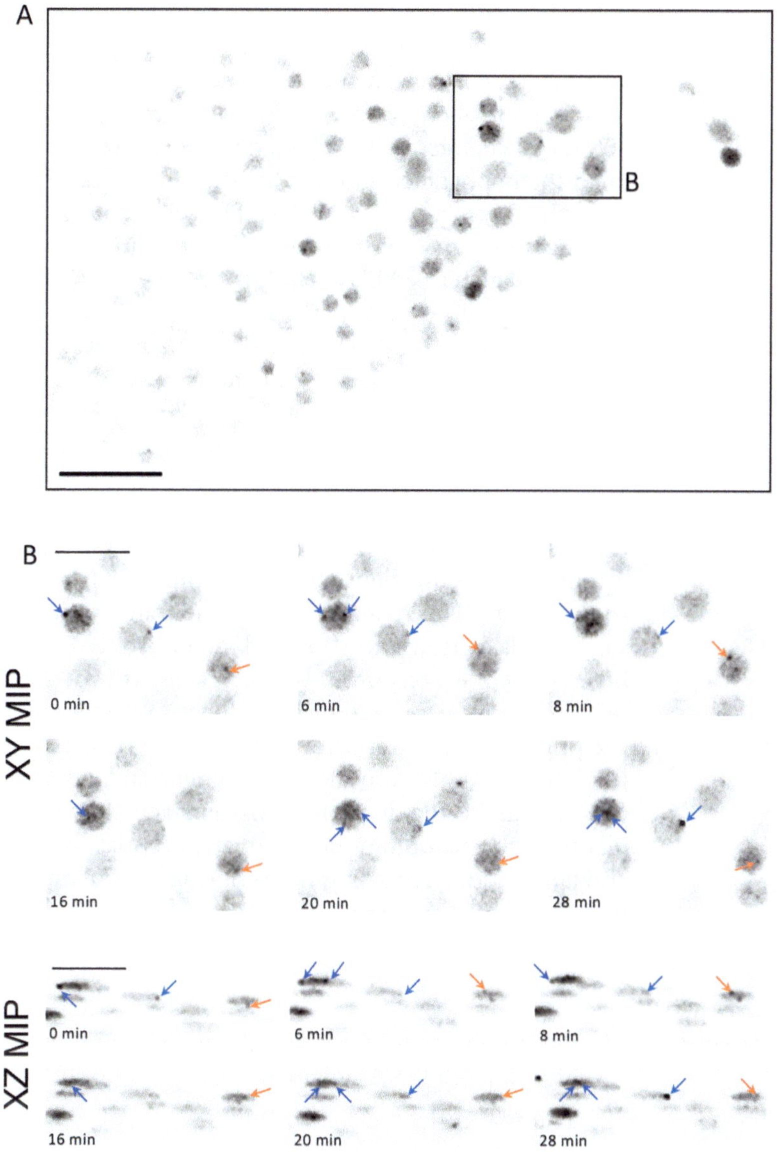

Fig. 3 Real-time visualization of transcription using the MS2 system reveals *SPX1* gene bursting at steady state, in phosphate deficiency. (**a**) Maximum intensity projection (MIP) of a time-lapse video recorded in 3D (66 z planes 600 nM apart), which displays MCP-GFP fluorescence in the transgenic plants. Scale bar, 20 μm. (**b**) (**c**) Maximum intensity projections on XY (**b**) or XZ (**c**) axes from the time-lapse video shown in (**a**) at the indicated time points and displaying MCP-GFP fluorescence. The arrows pinpoint transcription sites where the signal remains constant (orange) or switches between on/off states (blue) during the time lapse. Scale bars, 10 μm

dynamics in single cells in response to experimental treatments [10, 11].

In the present work, we focused on an early phosphate responsive gene (*SPX1*) that has been shown to be induced in phosphate deficiency and quickly repressed upon resupply. Phosphate is a crucial macronutrient for plant growth and productivity. To withstand its scarcity and maintain homeostasis, plants have developed tightly controlled processes and a series of developmental and metabolic changes [20, 21]. Gene expression analyses have underlined the importance and rapidity of transcriptional control underlying these mechanisms [21–23]. However, much of the phosphate sensing and signalling routes are barely known, and studying fast responsive genes (such as *SPX1*) to this element's fluctuations may aid in deciphering these pathways.

We used *SPX1* promoter to create transcriptional fusion to express MS2x128 repeats which were subsequently bound by ~256 MCP proteins fused to GFP. The success of the technique was first validated by smFISH (single molecule fluorescence in situ hybridization) using MS2 specific probes [11]. For in vivo experiments, to monitor the repression of our gene of interest, we combined live-cell imaging with root microfluidics, which allowed fast modification of nutrient regimes, without causing handling stress to the plants. The system dissected the transcriptional responses of single cells within the root and demonstrated the rapidity and heterogeneity of responses between cells. It brought forth a faster reporter system for systemic phosphate response as compared to the previously used reporter gene assays (GUS, GFP, luciferase...) [6, 24, 25]. It also unveiled phenomena such as gene bursting or pulsing, due to the discontinuous gene activity, for the first time in plants. The technology holds great potential for new discoveries and applications regarding transcriptional regulation in plantae.

2 Materials

2.1 Making Plasmid Constructs

1. DNA polymerase for high-fidelity PCR amplification (Phusion high-fidelity DNA polymerase).
2. Thermal cycler.
3. Agarose gel electrophoresis system and a gel imaging system.
4. PCR product purification and clean-up kit (NucleoSpin Gel and PCR clean-up kit).
5. pMK-MS2x128-XbaI vector containing MS2 tandem repeats (128 loops) (*see* **Note 1**).
6. pDONR201 carrying *MCP-NLS* (*see* **Note 1**).
7. pUB-Dest vector containing N-terminal eGFP fusion tag (*see* **Note 2**).
8. Modular Cloning (MoClo) destination vectors and cloning reagents.

9. DNA restriction endonucleases and buffers according to the manufacturer:
   - *Eco*R V, *Bpi* I, *Pme* I, *Sma* I, *Xba* I, and *Bsa* I.
10. Alkaline phosphatase.
11. T4 DNA polymerase.
12. T4 DNA ligase.
13. LR clonase II enzyme mix and proteinase K for Gateway cloning.
14. Chemically competent *E. coli* cells for transformation (One Shot TOP 10 *E. coli* and S.O.C medium).
15. High efficiency *E. coli* cells for large plasmid cloning (NEB 10-beta competent *E. coli* and its outgrowth medium).
16. LB liquid medium: dissolve 25 g of high salt LB broth in 1 L distilled water, and adjust the pH to 7.2 with NaOH. Autoclave it.
17. LB-agar medium: dissolve 15 g of agar and 25 g of high salt LB broth in 1 L of distilled water, and adjust the pH to 7.2 with NaOH. Autoclave and pour into plates while the solution is warm with the suitable antibiotic(s).
18. Sterile round Petri dishes.
19. Sterile culture tubes.
20. Microtubes 1.5 mL, DNAse free.
21. Antibiotics sterilized by filtration through 0.22 μm pore size filters, 1000 × stock solutions:
   - 50 mg/mL of kanamycin monosulfate in water. (Working concentration: 50 μg/mL).
   - 100 mg/mL of spectinomycin dihydrochloride hydrate in water. (Working concentration: 100 μg/mL).
22. For bacterial colonies screening following MoClo cloning, 100× stock solutions:
   - X-gal solution: 20 mg/mL of X-gal in dimethylformamide (DMF).
   - Isopropyl β-d-1-thiogalactopyranoside (IPTG) solution: 0.1 M IPTG in water.
23. Incubator for agar plates at 37 °C.
24. Shaker for liquid culture tubes at 37 °C (200 rpm).
25. Centrifuge for liquid culture tubes (10,000 rpm).
26. DNA extraction kit.

2.2 *Agrobacterium Mediated Transfer*

1. *Agrobacterium tumefaciens* strain C58C1 competent for electro-transformation.
2. Electroporation system and cuvettes.
3. Incubator for agar plates at 28 °C.
4. Shaker for liquid culture tubes at 28 °C (100 rpm).
5. Antibiotics sterilized by filtration through 0.22 μm pore size filters, 1000× stock solutions:
   - 50 mg/mL of kanamycin in water. (Working concentration: 50 μg/mL).
   - 50 mg/mL of rifampicin in methanol. (Working concentration: 50 μg/mL).
6. Sterile round Petri dishes.
7. Spectrophotometer to measure absorbance at 600 nM.
8. Distilled water, sucrose, and Silwett L-77.
9. Plant growth chambers for *Arabidopsis* plants under a 16 h light/8 h dark regime with 25 °C/22 °C, respectively.
10. *Arabidopsis* plants with floral buds.
11. Hoagland/2 medium: add 5 g/L of sucrose to 20 mL of Hoagland stock solution (composition on Table 1). Add 8 g/L of agar. Autoclave it.
12. In vitro plant growth medium A: add 8 g of agar and 5 g sucrose to MES (0.68 g/L final concentration) and a nutritive solution [26]. Autoclave and pour into plates while the solution is warm.

Table 1
Hoagland/2 medium composition for 50× stock solution

Compound	Final concentration	
Mg SO_4, $7H_2O$	1	mM
$Ca(NO_3)2,4H_2O$	2	mM
KNO_3	1.63	mM
$NH_4H_2PO_4$	0.5	mM
Sequestren	1.6	μM
H_3BO_3	44.9	μM
$MnCl_2$, $4H_2O$	9	μM
$ZnSO_4$, $7H_2O$	0.8	μM
$CuSO_4$, $5H_2O$	0.3	μM
Na_2MoO_4, $2H_2O$	1.2	μM

13. KH_2PO_4 1 M in water solution.
14. Microscopy slides and coverslips.
15. Laser scanning confocal microscope with specific filtering for GFP and high magnification objectives (40× (water) and 63× (oil)).

2.3 Root Microfluidics

1. Polydimethylsiloxane (PDMS) RootChips (8S) fixed on a large glass coverslip [27, 28].
2. In vitro plant growth medium B with or without agar (composition on Table 2): add 1 g/L of MES hydrate. Adjust pH to 5.7 with KOH. For solid media preparation, add 8 g/L agar. Autoclave medium then add 2 μM $FeCl_2$.
3. Pipette tips.
4. Razor blade.
5. Micropore tape.
6. Tubes (0.02 I.D.x 0.06 O.D.x0.02 Wall) and connectors.
7. Multi-channel microfluidic pump.
8. Autoclaved ultrapure water.
9. Ethanol 70%.
10. 3D printed plastic carrier for the RootChip, adapted to the microscope stage.
11. Sterilized cotton (to imbibe with water and create humidity for the plants in the RootChip).

Table 2
Plant growth medium (B) composition for 1 L (adapted from MS medium)

Compound	Concentration (mg/L)
$MgSO_4.7H_2O$	185
NH_4NO_3	825
KNO_3	950
$CaCl_2.2H_2O$	220
KI	0.415
H_3BO_3	3.1
$MnSO_4.H_2O$	8.45
$ZnSO_4.7H_2O$	4.3
$Na_2MoO_4.2H_2O$	0.125
$CuSO_4.5H_2O$	0.0125
$CoCl_2.6H_2O$	0.0125

2.4 Microscopy and Live-Cell Imaging

1. Spinning disk mounted with a Nikon Eclipse Ti, equipped with specific filtering for GFP and SynaptoRed, and a x40 Plan-Apo water objective (1.15 NA; 0.6 mM DT).
2. In vitro plant growth medium B (without agar) supplied with 500 μM of KH_2PO_4.
3. SyntaptoRed C2.
4. Microscope image acquisition and analysis softwares (ImageJ).

3 Methods

3.1 Cloning of Plasmid Constructs

Imaging RNA with the MS2-MCP system involves the co-expression of MS2-tagged RNA together with a fluorescently tagged MCP.

To make these two constructs, we used a combination of modular cloning technique (MoClo) [29] and Gateway technologies (*see* **Note 3**).

3.1.1 Fusion of 128 MS2 Loops to the Promotor of Interest (pOI)

First for the reporter mRNA construct, we describe the fusion of 128*MS2 loops to our promotor of interest by MoClo. MoClo allows to assemble multiple genetic elements in various predefined arrangements. The choice of vector for each module can be adapted according to the desired final construct. We made a transcriptional fusion using the promoter of interest (pOI) and inserted the MS2*128 right after its 5′UTR (*see* **Note 4**).

1. Use high-fidelity polymerase and specific primers to amplify the 1.5 kb region upstream of the start codon of the gene of interest. The primer sequences should contain *Bpi* I recognition sites at the outer extremities for further cloning.
2. Use a small fraction of the PCR product to verify the expected size of the amplified DNA fragment by agarose gel electrophoresis, and purify the rest using a PCR purification kit following manufacturer's instructions.
3. Insert the purified PCR product in an appropriate level 0 plasmid (pICH41295, *see* **Note 5**) in a 2:1 insert-to-vector ratio using 100 ng of vector. Follow the classical MoClo cloning protocol [29] (*see* **Note 6**; Fig. 4).
4. Due to the repetitive MS2 sequences and difficulty of amplification by PCR, classical cloning strategy using type I restriction enzymes is adopted.

   Excise the MS2x128 from pMK-MS2x128-XbaI (*see* **Note 1**) using blunt end cutters *Sma*I and *EcoR*V. Cut the level 0 plasmid (pICH41308, *see* **Note 5**) open using *Bpi*I to remove the LacZ gene, and fill the created overhangs with nucleotides using T4 DNA polymerase. Dephosphorylate the vector using

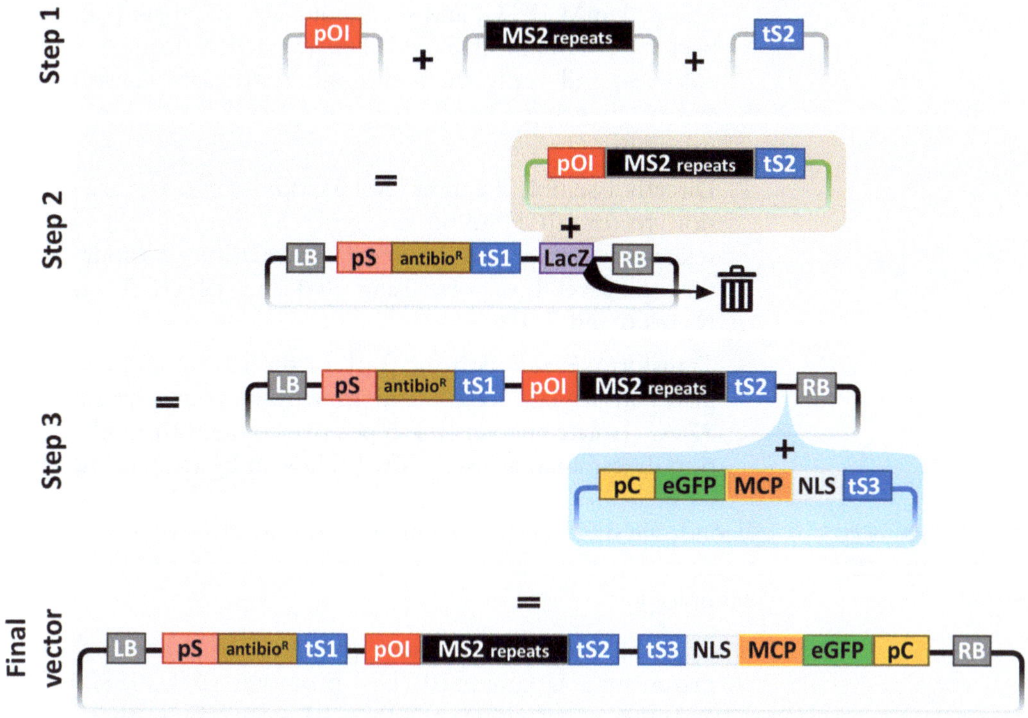

Fig. 4 Diagram showing the cloning steps of the vector aimed at introducing the MCP-MS2 system in *Arabidopsis*. **Step 1**: Assembly of the expression unit of MS2 repeat expression unit by MoClo technique. Each component of the transcription unit (promotor of interest (pOI), MS2 repeats, terminator (tS2)) was previously introduced into a "Level 0" vector according to the MoClo protocol. **Step 2**: Replacement of the LacZ gene of the final destination vector with MS2 repeat expression cassette. Note that the final destination vector already carries an expression cassette of a resistance gene (antibio^R), under the control of a strong promoter (pS) and has a strong terminator (tS1). **Step 3**: Completion of the final vector by introducing the expression cassette including the MCP-GFP-NLS coding sequence under the control of a constitutive promotor (pC) and a strong terminator (tS3). The *final vector* thus comprises three expression cassettes allowing the expression of MCP coupled to GFP, an RNA carrying MS2 repeats, and conferring antibiotic resistance to the plant. These three cassettes are located between the left border (LB) and right border (RB) regions

alkaline phosphatase to prevent its self-ligation. Ligate the insert and vector (100 ng) in a 7:1 ratio using DNA ligase (Fig. 4).

5. Use the level 0 plasmid pICH41421 carrying a Nopaline synthase terminator (*see* **Note 5**) (Fig. 4).
6. All level 0 destination vectors confer spectinomycin resistance and encode a lacZα fragment for blue/white screening. After each of the steps 3 and 4, transform competent *E. coli* cells with 3 μL of the ligation mix.

   Use One Shot TOP 10 chemically competent *E. coli* for the promoter and terminator modules and high-efficiency NEB 10-beta for the MS2.

Plate bacteria on Petri dishes supplied with 0.2 mg/mL X-Gal, 1 mM IPTG, and spectinomycin. Incubate plates with bacteria overnight at 37 °C. Isolate bacteria from 4–5 white colonies, and verify them through restriction digest analysis and sequencing. Verify that the MS2 is in the correct orientation.

7. Directly assemble the three level 0 modules into a level 2 expression vector pICH86966 (*see* **Note 5**) carrying a kanamycin selection marker for plant transformation without going through level 1 vectors, using *Bsa*I and T4 DNA ligase (*see* **Notes 6** and 7; Fig. 4).
8. Transform *E. coli* with the 3 μL of the level 2 construct, and plate on X-Gal/ IPTG and kanamycin plates overnight at 37 °C. Isolate and verify 4–5 positive colonies through restriction digest analysis using *Xba* I followed by sequencing.

3.1.2 Assembly of the MS2 Coat Protein (MCP) and GFP

Second, we describe the assembly of the MCP with a fluorescent protein under the control of a constitutive promoter using Gateway technology.

1. Perform LR reaction using pUBN-Dest containing eGFP driven by a Ubiquitin10 (Ub) promoter (At4G05320) [30] and the entry clone pDONR201 containing MCP and a nuclear localization signal (NLS) (*see* **Note 1**). Mix 300 ng of pUBN-Dest vector with 300 ng of the entry clone, and use the LR clonase II enzyme mix following the manufacturer's instructions.
2. Transform *E. coli* with 1 μL of the ligation, and plate on LB media agar plates supplied with spectinomycin for selection. Isolate positive colonies, and verify the clones through restriction digest analysis.

3.1.3 Final T-DNA Construct

Third, we describe the insertion of the two constructs onto the same T-DNA.

1. Cut the pOI::MS2x128 vector using *Pme*I, and dephosphorylate it using alkaline phosphatase to prevent its self-ligation.
2. Excise the p*Ub::eGFP/MCP-NLS* from the Gateway vector using *Sma*I and *Pme*I, and introduce it into the pOI:: MS2x128 vector. Use a 5:1 insert to vector ratio and a T4 DNA ligase to ligate the two (Fig. 4).
3. Transform NEB-10 beta competent *E. coli* cells with 3 μL of the ligation, and plate them on LB Petri dishes supplied with kanamycin (*see* **Note 8**). Incubate them overnight at 37 °C.
4. Isolate 4–5 positive colonies, and verify them through PCR using GFP specific primers, followed by sequencing.

3.2 *Agrobacterium-Mediated Transfer of the MS2-MCP Plasmid in* Arabidopsis thaliana

1. Transform electrocompetent *Agrobacterium tumefaciens* strain C58C1 with the obtained positive clone carrying the complete construct by electroporation. Plate them on LB agar plates containing rifampicin and kanamycin, and incubate them for 48 h at 28 °C.
2. Start 4–5 liquid cultures from isolated bacterial colonies in tubes containing 10 mL LB medium under the same antibiotic selection as the previous step. Incubate the colonies 48 h at 28 ° C with agitation (200 rpm). Extract the plasmids, and verify the presence of the construct by PCR and gel electrophoreses.
3. Grow the positive colony again in liquid LB culture supplied with rifampicin and kanamycin for 48 h at 28 °C. Transfer 3 mL of the culture into a tube and supply it with 27 mL of LB. Let it grow at 28 °C for about 5 h, and then measure the optical density (OD) of the bacteria. When it reaches approximately 1 OD, supply the culture with 120 mL water, 6 g of sucrose, and 36 μL of Silwett L-77.
4. Transform *Arabidopsis thaliana* with the bacterial solution via floral dipping.
5. Select the obtained transformants on Hoagland/2 media supplied with kanamycin. Carry the progeny exhibiting 3:1 segregation to T3 generation to identify homozygous lines carrying a single insertion locus.
6. To verify that the RNA-tagging system is functional in the obtained lines, grow some seeds in conditions that would normally induce the MS2-tagged gene and others in repressing conditions as a control. Mount the plants on a microscopic slide in appropriate liquid media.
7. Image the plants on a confocal microscope (Zeiss LSM 780) using a 488 nM laser. Use a high-magnification objective to visualize the presence of the MS2-tagged RNA labelled by MCP-GFP in the nuclei of MS2-induced plants. Select the lines that display a clear signal with minimal fluorescent background. Specific fluorescent signals should be absent in the control samples that were grown in non-inducing conditions.
8. For additional confirmation of the MS2-RNA tagging, perform smFISH using probes against the MS2 stem loops (*see* **Note 9**).

3.3 Root Microfluidics

We describe next the setup of the root microfluidic system [31] that allows to grow *Arabidopsis* seedlings in a polydimethylsiloxane (PDMS)-based RootChip-8S [27, 28]. The RootChip is composed of eight channels coupled to a perfusion system based on a multichannel microliter pump. This allows to feed the plantlets and control the root microenvironment precisely and rapidly. This method provides the means to combine live-cell imaging with the capacity to change the nutritive solution without causing any stress to the plants.

3.3.1 Preparation of Plastic Cones and Seedling Growth

1. Fill a Petri dish with in vitro plant growth medium B, suitable for the experimental requirements. Avoid adding sucrose to minimize microbial contamination.
2. While the medium has not set yet, use a multi-channel pipette to fill pipette tips with 10 μL of the medium. They must be of an appropriate diameter size to fit the RootChip tightly and prevent leaks (Fig. 5a).
3. Once the medium in the tips is set, cut to 4–5-mM-long cones, and place them upright in a Petri dish containing solid growth medium (Fig. 5b).
4. Place a single *Arabidopsis* seed on top of each cone. Add 1 mL of sterilized water on the agar to create humidity in the Petri dish (Fig. 5b).
5. Seal the Petri dish with two rounds of micropore tape, and store overnight at 4 °C for stratification.
6. Transfer the plates to a growth cabinet with 16 h light/8 h dark regime with 25 °C/22 °C, respectively.
7. The seedlings should be ready for transfer to the RootChip within 5–7 days. Under a microscope, select individual seedlings whose root tip is near the bottom outlet of the plastic cone.

3.3.2 Transfer of Seedlings onto the RootChip

1. Plug the tubing connectors onto the RootChip (Fig. 6). A multi-channel microliter pump is used to control the flow of liquid into chip. Flush the tubes and channels with 70% ethanol then by several rounds of sterilized water followed by several rounds of liquid growth media.
2. Plug the plastic cones with seedling that were selected in III. A.7 into the RootChip inlets (Fig. 5c). They should fit tightly, but don't apply too much pressure to the chip not to break the optical glass.
3. Fit the RootChip onto a carrier that was designed and printed to fit the microscope stage. Imbibe sterilize cotton with autoclaved water, and place them in the reservoirs of the carrier to maintain appropriate humidity and keep the plants from drying out. Seal the cover of the plastic carrier with micropore tape (Fig. 6).
4. Set the liquid media flow speed to 5 rpm. Depending on the desired flow rate, the pressure may be adjusted.
5. Transfer the carrier into the growth chamber. Place it at a 45 ° angle to facilitate the entry of the roots into channels, and monitor plant growth (*see* **Note 10**).

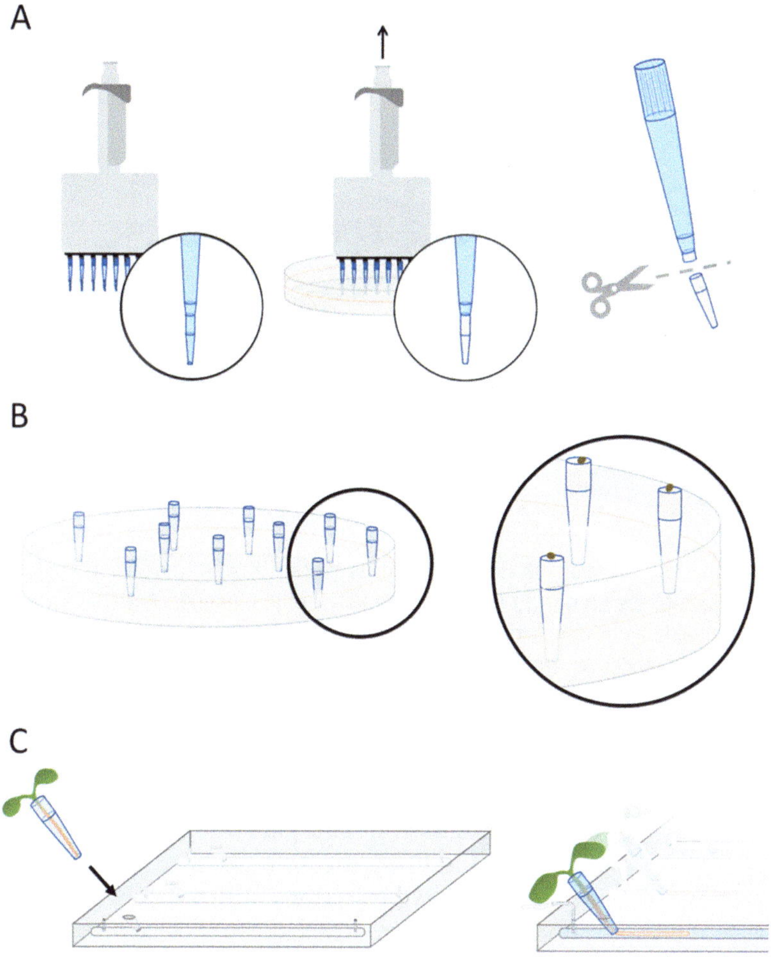

Fig. 5 Preculture of plants on agar plate before their insertion on RootChip. (**a**) Using a multichannel pipette, plastic cone tips are filled up with melted agar. When the agar is solidified, the tip of the cone is cut off with scissors or a razor blade. (**b**) Cone tips are inserted vertically onto an agar plate. One seed is deposited on the agar at the top of each cone. (**c**) Few days after germination, when the root apex reaches the lower end of the cone, the latter is transferred on a RootChip

3.3.3 Microscopy and Live-Cell Imaging

1. Once the roots reach the desired length in the RootChip, transfer the carrier, and fit it into the microscope stage. Place the pump and the liquid solutions on a table nearby to avoid vibrations of the microscope.
2. Perform in vivo imaging using an Andor Dragonfly spinning disk mounted with a Nikon Eclipse Ti using a x40 Plan-Apo water objective (1.15 NA; 0.6 mM DT). In our experience, this microscope offers the best compromise between image quality, speed of acquisition, and size of the field of view.

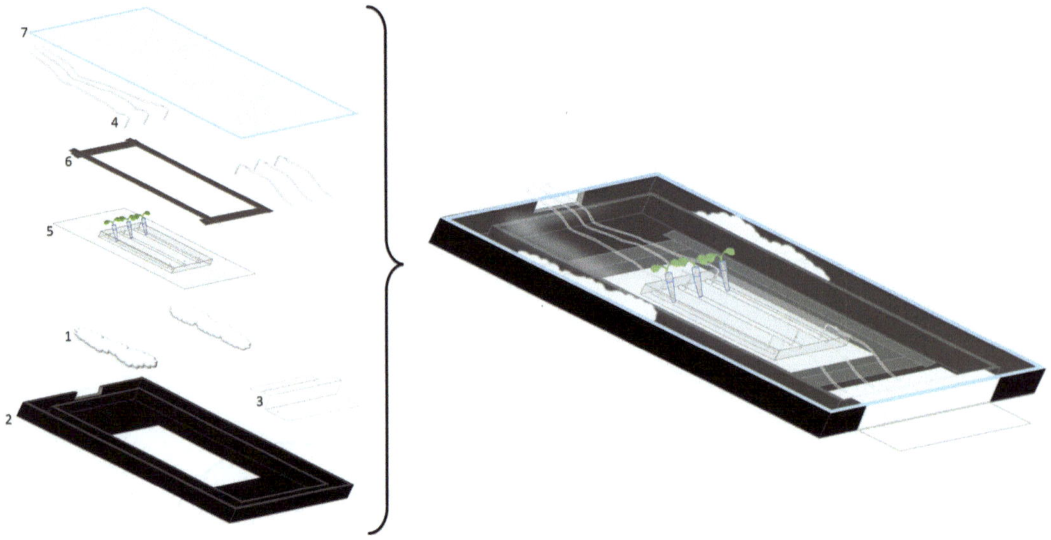

Fig. 6 Diagram of the assembly of the RootChip and its carrier. Sterile, moist cotton wool pieces (**1**) are wedged into the ledges of the carrier (**2**) to create humidity in the chamber. A sterile Whatman paper (**3**) is placed between the inside and the outside of the chamber to drain off any overflow of liquid from the tubes. Tubes that convey liquid media (**4**) are connected to the inlet and outlet of the root chip (**5**). The seedlings, having grown in the cones, are plugged on the RootChip which is then fixed on the carrier by adhesive tape (**6**). The chamber is closed by a transparent Plexiglass plate (**7**) which is sealed to the carrier with micropore tape

3. Acquire time-lapse videos, making Z-stacks every 2–3 min with a scan size of 100–150 μm (0.5 μm step size; ~200 steps per z-stack) and using the 488 nM laser for about an hour. Maximum intensity projections of the images produced show multiple transcription sites that could be later analyzed and confirm the transcriptional heterogeneity between cells. These observations of the root in steady-state conditions allow to study events such as transcriptional bursting where the transcription switches between on and off states. Putative single RNA molecules could also be seen and studied. The system also granted access to the different root cell layers.
4. In order to study events such as transcriptional induction or repression of the MS2-tagged gene which is related to nutritional stress, change the nutritive solution of the plant that is being imaged. Add an inert fluorescent dye to the changed nutritive solution, such as 5 μg/mL SynaptoRed for instance, to precisely record the arrival of the media to the imaging field and accurately define the time course. Acquire time-lapse videos with the same settings as in III.C.3, using a 488 nM laser and one adapted to the chosen inert fluorescent dye.
5. After acquisition of the time-lapse movies, perform a 2D maximum intensity projection.

6. Use ImageJ and the TrackMate plug-in or similar software (e.g., Imaris) to analyze transcription dynamics. TrackMate has a user-friendly interface that allows tracking of the signal (here transcription site dynamics) during the time course experiments [35, 36].

   (In our case: Assess the intensity of the transcription site for several nuclei per sample. In each nucleus, subtract the mean intensity of the nuclear background from the maximum intensity of the transcription site and plot the resulting intensities against time).

4 Notes

1. pMK-MS2x128-XbaI vector containing 128xMS2 loops and the entry clone pDONR201 containing MCP and a nuclear localization signal (NLS) were obtained from E. BERTRAND and can be provided upon request.

   For the MCP we used a deltaFG-V29I variant. V29I is a mutation that grants a fivefold increase in the affinity of the MCP, and deltaFG is a deletion that prevents multimerization of the MCP into a capsid protein but still allows its dimerization.
2. pUB-Dest vector containing N-terminal eGFP fusion tag was provided from V. RUBIO (CNB- CSIC, Madrid).
3. Alternatively, to the proposed cloning strategy, one could synthesize the constructs based on MoClo solely, and insert the MS2 elsewhere on the gene of interest. Complex constructs containing many transcription units can be assembled by a series of cloning reactions with high efficiency. The technique offers great flexibility in the design of cassettes and the possibility of reusing each of the modules in different genetic constructs. Over the past few years, plant-specific MoClo modules have been developed to widen its use. We now fully adopt it to make our MS2-tagged RNA constructs (Fig. 7).
4. The MoClo modules could be designed to include the coding sequence of the gene of interest, and the MS2 can be inserted in other areas of the gene depending on the purpose of the study (in an intron, 3′UTR) [13, 34]. Here we only made a transcriptional fusion using the gene's promoter to visualize transcription.
5. All the "pICH" vectors come from MoClo plasmid bank [32, 33].
6. If *Bpi*I or *Bsa*I restriction enzymes cut inside the sequence being cloned, modify the digestion-ligation protocol by increasing the insert-to-vector ratio, and eliminate the last

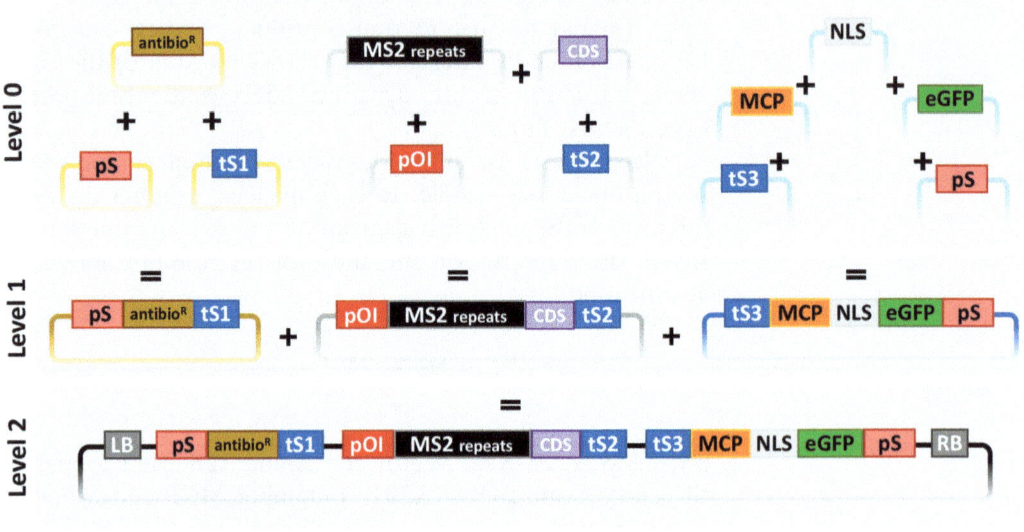

Fig. 7 Diagram showing an alternative way of cloning using only MoClo system. *Level 0*: The different modules are inserted in level 0 plasmids or taken from a vector database (ref??) to make a complete transcription unit. *Level 1*: Combine the different level 0 modules to create transcription units via classical MoClo protocol. *Level 2*: Combine the different transcription units to create a multigene construct. Promotor of interest (pOI), terminator (tS), expression cassette of a resistance gene (antibio^R), coding sequence of your gene of interest (CDS), constitutive promotor (pC)

Table 3
Shortened PCR cycle to avoid unwanted cut-off by *BpiI* and/or *BsaI*

Step	Number of cycles	T (°C)	Time (min)
1	5	37	5
		16	10
2	1	80	5

15 min at 37 °C step which would normally allow the removal of the empty vector (Table 3). This increases the efficiency of the cloning and prevents the loss of any gene fragment.

7. We tested both the long and short PCR programs proposed for the MoClo technique. (https://synbio.tsl.ac.uk/uploads/a94d80d88a4d03acbe8b53da74fdcf49.pdf). For our cloning, we found that the long cycle provided higher cloning efficiency (Table 4).
8. Alternatively, Stbl2 or Stbl4 (Invitrogen) could be used as they have high transformation efficiency and are suitable for cloning unstable inserts and larger constructs.

Table 4
Long PCR cycle

Step	Number of cycles	T (°C)	Time (min)
1	26	37	3
		16	4
2	1	50	15
3	1	80	5

9. The detailed smFISH protocol along with the probe sequences targeting the MS2 repeats can be found in [11]. The image analysis and computation methods were also described.
10. In our hands, plants in the RootChips grew better in a chamber with minimal air flow which maintained favorable humidity.

Acknowledgments

SH was supported by a PhD fellowship from the CEA and PACA region and CM by the French National Research Agency (ANR-18-CE20-023) and the Centre Mondial de l'Innovation-Roullier Group (CM, CIFRE grant number 2018/1078). Additional grant support was also received by ANR ULTIM (21-CE-120016) and PHLOWZ (19-CE-13-0007). The pMK-MS2x128-XbaI vector containing 128xMS2 loops and the entry clone pDONR201 containing MCP and a nuclear localization signal (NLS) were provided by EB. pUB-Dest vector containing N-terminal eGFP fusion tag was provided by V. RUBIO (CNB-CSIC, Madrid). We acknowledge the MRI imaging facility (belonging to the National Infrastructure France-BioImaging supported by the French National Research Agency, ANR-10-INBS-04). We thank H. JAVOT for her help in setting up the microfluidic system, G. GROSSMANN for providing the RootChips-8S, and M. PHILIBERT for the 3D printing of the carriers and the ZoOM microscopy platform (supported by the Région Provence Alpes Côte d'Azur, the Conseil General of Bouches du Rhône, the French Ministry of Research, the Centre National de la Recherche Scientifique, and the Commissariat à l'Energie Atomique et aux Energies Alternatives). The original design for the 3D printing of carriers (compatible with LSM780 stages) was provided by D. BOUCHEZ and was further modified by H. JAVOT and S. SAUGE-MERLE.

References

1. Baillo K, Zhang X (2019) Transcription factors associated with abiotic and biotic stress tolerance and their potential for crops improvement. Genes 10:771. https://doi.org/10.3390/genes10100771
2. Vaquerizas JM, Kummerfeld SK, Teichmann SA, Luscombe NM (2009) A census of human transcription factors: function, expression and evolution. Nat Rev Genet 10:252–263. https://doi.org/10.1038/nrg2538
3. Klimyuk VI, Nussaume L, Harrison K, Jones JDG (1995) Novel GUS expression patterns following transposition of an enhancer trap Ds element in Arabidopsis. Mol Gen Genet 9
4. Krouk G, Crawford NM, Coruzzi GM, Tsay Y-F (2010) Nitrate signaling: adaptation to fluctuating environments. Curr Opin Plant Biol 13:265–272. https://doi.org/10.1016/j.pbi.2009.12.003
5. Crisp PA, Ganguly DR, Smith AB, Murray KD, Estavillo GM, Searle I, Ford E, Bogdanović O, Lister R, Borevitz JO, Eichten SR, Pogson BJ (2017) Rapid recovery gene downregulation during excess-light stress and recovery in arabidopsis. Plant Cell 29:1836–1863. https://doi.org/10.1105/tpc.16.00828
6. Hanchi M, Thibaud M-C, Légeret B, Kuwata K, Pochon N, Beisson F, Cao A, Cuyas L, David P, Doerner P, Ferjani A, Lai F, Li-Beisson Y, Mutterer J, Philibert M, Raghothama KG, Rivasseau C, Secco D, Whelan J, Nussaume L, Javot H (2018) The phosphate fast-responsive genes *PECP1* and *PPsPase1* affect phosphocholine and phosphoethanolamine content. Plant Physiol 176: 2943–2962. https://doi.org/10.1104/pp.17.01246
7. Bertrand E, Chartrand P, Schaefer M, Shenoy SM, Singer RH, Long RM (1998) Localization of ASH1 mRNA particles in living yeast. Mol Cell 2:437–445. https://doi.org/10.1016/S1097-2765(00)80143-4
8. Schmidt U, Basyuk E, Robert M-C, Yoshida M, Villemin J-P, Auboeuf D, Aitken S, Bertrand E (2011) Real-time imaging of cotranscriptional splicing reveals a kinetic model that reduces noise: implications for alternative splicing regulation. J Cell Biol 193:819–829. https://doi.org/10.1083/jcb.201009012
9. Horvathova I, Voigt F, Kotrys AV, Zhan Y, Artus-Revel CG, Eglinger J, Stadler MB, Giorgetti L, Chao JA (2017) The dynamics of mRNA turnover revealed by single-molecule imaging in single cells. Mol Cell 68:615–625. e9. https://doi.org/10.1016/j.molcel.2017.09.030
10. Alamos S, Reimer A, Niyogi KK, Garcia HG (2021) Quantitative imaging of RNA polymerase II activity in plants reveals the single-cell basis of tissue-wide transcriptional dynamics. Nat Plants 36
11. Hani S, Cuyas L, David P, Secco D, Whelan J, Thibaud M-C, Merret R, Mueller F, Pochon N, Javot H, Faklaris O, Maréchal E, Bertrand E, Nussaume L (2021) Live single-cell transcriptional dynamics via RNA labelling during the phosphate response in plants. Nat Plants 7:1050–1064. https://doi.org/10.1038/s41477-021-00981-3
12. Ferraro T, Esposito E, Mancini L, Ng S, Lucas T, Coppey M, Dostatni N, Walczak AM, Levine M, Lagha M (2016) Transcriptional memory in the drosophila embryo. Curr Biol 26:212–218. https://doi.org/10.1016/j.cub.2015.11.058
13. Tantale K, Mueller F, Kozulic-Pirher A, Lesne A, Victor J-M, Robert M-C, Capozi S, Chouaib R, Bäcker V, Mateos-Langerak J, Darzacq X, Zimmer C, Basyuk E, Bertrand E (2016) A single-molecule view of transcription reveals convoys of RNA polymerases and multiscale bursting. Nat Commun 7:12248. https://doi.org/10.1038/ncomms12248
14. Lammers NC, Kim YJ, Zhao J, Garcia HG (2020) A matter of time: using dynamics and theory to uncover mechanisms of transcriptional bursting. Curr Opin Cell Biol 67:147–157. https://doi.org/10.1016/j.ceb.2020.08.001
15. Zhang G, Simon AE (2003) A multifunctional turnip crinkle virus replication enhancer revealed by in vivo functional SELEX. J Mol Biol 326:35–48. https://doi.org/10.1016/S0022-2836(02)01366-9
16. Hamada S, Ishiyama K, Choi S-B, Wang C, Singh S, Kawai N, Franceschi VR, Okita TW (2003) The transport of prolamine RNAs to prolamine protein bodies in living rice endosperm cells [W]. Plant Cell 15:2253–2264. https://doi.org/10.1105/tpc.013466
17. Fang Y, Spector DL (2007) Identification of nuclear dicing bodies containing proteins for MicroRNA biogenesis in living arabidopsis plants. Curr Biol 17:818–823. https://doi.org/10.1016/j.cub.2007.04.005
18. Sambade A, Brandner K, Hofmann C, Seemanpillai M, Mutterer J, Heinlein M (2008) Transport of TMV movement protein particles associated with the targeting of RNA

to plasmodesmata. Traffic 9:2073–2088. https://doi.org/10.1111/j.1600-0854.2008.00824.x

19. Fujioka Y, Utsumi M, Ohba Y, Watanabe Y (2007) Location of a possible miRNA processing site in SmD3/SmB nuclear bodies in arabidopsis. Plant Cell Physiol 48:1243–1253. https://doi.org/10.1093/pcp/pcm099
20. Rouached H, Arpat AB, Poirier Y (2010) Regulation of phosphate starvation responses in plants: signaling players and cross-talks. Mol Plant 3:288–299. https://doi.org/10.1093/mp/ssp120
21. Misson J, Raghothama KG, Jain A, Jouhet J, Block MA, Bligny R, Ortet P, Creff A, Somerville S, Rolland N, Doumas P, Nacry P, Herrerra-Estrella L, Nussaume L, Thibaud M-C (2005) A genome-wide transcriptional analysis using Arabidopsis thaliana Affymetrix gene chips determined plant responses to phosphate deprivation. Proc Natl Acad Sci 102:11934–11939. https://doi.org/10.1073/pnas.0505266102
22. Barragán-Rosillo AC, Peralta-Alvarez CA, Ojeda-Rivera JO, Arzate-Mejía RG, Recillas-Targa F, Herrera-Estrella L (2021) Genome accessibility dynamics in response to phosphate limitation is controlled by the PHR1 family of transcription factors in *Arabidopsis*. Proc Natl Acad Sci USA 118:e2107558118. https://doi.org/10.1073/pnas.2107558118
23. Thibaud M-C, Arrighi J-F, Bayle V, Chiarenza S, Creff A, Bustos R, Paz-Ares J, Poirier Y, Nussaume L (2010) Dissection of local and systemic transcriptional responses to phosphate starvation in Arabidopsis: local versus systemic responses to phosphate starvation. Plant J 64:775–789. https://doi.org/10.1111/j.1365-313X.2010.04375.x
24. Kanno S, Cuyas L, Javot H, Bligny R, Gout E, Dartevelle T, Hanchi M, Nakanishi TM, Thibaud M-C, Nussaume L (2016) Performance and limitations of phosphate quantification: guidelines for plant biologists. Plant Cell Physiol 57:690–706. https://doi.org/10.1093/pcp/pcv208
25. Misson J, Thibaud M-C, Bechtold N, Raghothama K, Nussaume L (2004) Transcriptional regulation and functional properties of Arabidopsis Pht1;4, a high affinity transporter contributing greatly to phosphate uptake in phosphate deprived plants. Plant Mol Biol 55:727–741. https://doi.org/10.1007/s11103-004-1965-5
26. Mercier C, Roux B, Have M, Le Poder L, Duong N, David P, Leonhardt N, Blanchard L, Naumann C, Abel S, Cuyas L, Pluchon S, Nussaume L, Desnos T (2021) Root responses to aluminium and iron stresses require the SIZ1 SUMO ligase to modulate the STOP1 transcription factor. Plant J:15525. https://doi.org/10.1111/tpj.15525
27. Denninger P, Reichelt A, Schmidt VAF, Mehlhorn DG, Asseck LY, Stanley CE, Keinath NF, Evers J-F, Grefen C, Grossmann G (2019) Distinct RopGEFs successively drive polarization and outgrowth of root hairs. Curr Biol 29:1854–1865.e5. https://doi.org/10.1016/j.cub.2019.04.059
28. Guichard M, Bertran Garcia de Olalla E, Stanley CE, Grossmann G (2020) Microfluidic systems for plant root imaging. In: Methods in cell biology. Elsevier, pp 381–404
29. Weber E, Engler C, Gruetzner R, Werner S, Marillonnet S (2011) A modular cloning system for standardized assembly of multigene constructs. PLoS One 6:e16765. https://doi.org/10.1371/journal.pone.0016765
30. Grefen C, Donald N, Hashimoto K, Kudla J, Schumacher K, Blatt MR (2010) A ubiquitin-10 promoter-based vector set for fluorescent protein tagging facilitates temporal stability and native protein distribution in transient and stable expression studies: fluorescence tagging and expression in Arabidopsis. Plant J 64:355–365. https://doi.org/10.1111/j.1365-313X.2010.04322.x
31. Deal RB, Henikoff S (2011) The INTACT method for cell type–specific gene expression and chromatin profiling in Arabidopsis thaliana. Nat Protoc 6:56–68. https://doi.org/10.1038/nprot.2010.175
32. Engler C, Youles M, Gruetzner R, Ehnert T-M, Werner S, Jones JDG, Patron NJ, Marillonnet S (2014) A golden gate modular cloning toolbox for plants. ACS Synth Biol 3:839–843. https://doi.org/10.1021/sb4001504
33. Gantner J, Ordon J, Ilse T, Kretschmer C, Gruetzner R, Löfke C, Dagdas Y, Bürstenbinder K, Marillonnet S, Stuttmann J (2018) Peripheral infrastructure vectors and an extended set of plant parts for the modular

cloning system. PLoS One 13:e0197185. https://doi.org/10.1371/journal.pone.0197185

34. Pichon X, Lagha M, Mueller F, Bertrand E (2018) A growing toolbox to image gene expression in single cells: sensitive approaches for demanding challenges. Mol Cell 71(3): 468–480
35. Ershov D, Phan MS, Pylvänäinen JW, Rigaud SU, Le Blanc L, Charles-Orszag A, Conway JR, Laine RF, Roy NH, Bonazzi D, Duménil G (2021) Bringing TrackMate in the era of machine-learning and deep-learning. bioRxiv
36. Tinevez JY, Perry N, Schindelin J, Hoopes GM, Reynolds GD, Laplantine E, Bednarek SY, Shorte SL, Eliceiri KW (2017) TrackMate: an open and extensible platform for single-particle tracking. Methods 115:80–90

Chapter 5

Live-Cell RNA Imaging with a DNA-Functionalized Metal-Organic Framework-Based Fluorescent Probe

Hongyan Zhang and Libing Zhang

Abstract

Live-cell fluorescence imaging of tumor-associated miRNAs is significant for understanding the cancer onset and progression and the diagnosis and prognosis of clinical diseases. In this protocol, we describe the construction of a DNA-functionalized metal-organic framework-based fluorescent probe and demonstrate that the probe can be used to detect miRNA in vitro and live-cell imaging. The formed DNA-MOF probe can specifically target miRNA with high sensitivity and specificity. The detection limit of the extracellular assay is as low as the picomolar level. In addition, we found that the probe could permeate cells and can be successfully applied in intracellular miRNA imaging, even achieving the distinction of cancer cells and normal cells, as well as the cancer cell lines with different miRNA expression levels.

Key words DNA, Metal-organic framework, miRNA, Fluorescence probe, Imaging

1 Introduction

MicroRNA (miRNA) is a class of short non-coding RNA that plays a pivotal role in regulating gene expression in the transcription stage [1]. It widely exists in eukaryotic cells and functions as regulating cell proliferation, differentiation, and apoptosis [2]. It has been proven that the expression dysregulation of the miRNA is inseparable from tumorigenesis which makes it imperative to monitor and image the expression level and distribution of miRNA [3, 4]. Traditional detection strategies including northern blotting, quantitative reverse transcription-polymerase chain reaction (qRT-PCR), and microarrays have been used for the detection of miRNA. However, those methods still face the limitations of cumbersome operations, complicated primer design processes, and dependence on specialized equipment [5]. Therefore, there is a pressing demand to develop new strategies with high selectivity and precision to monitor and track extracellular and intracellular miRNAs [6].

Shin-ichi Sato (ed.), *Live-Cell RNA Imaging: Methods and Protocols*, Methods in Molecular Biology, vol. 2875,
https://doi.org/10.1007/978-1-0716-4248-1_5,

Recently, fluorescence methods have emerged as a potential candidate to compensate for the defects of the aforementioned methods and to be employed for in situ imaging of intracellular miRNAs due to their simplicity of operation and high sensitivity [7–9]. Especially, compared with the single-signal fluorescent probes, ratiometric fluorescent probes can eliminate the effect of the external environment and probe concentration by utilizing the ratio of the two output signals, thereby improving the detection sensitivity and accuracy [10, 11]. Benefit from the large specific surface area and good biocompatibility of the nanomaterials, the incorporation of nanomaterials to prepare DNA-functionalized nanomaterial-based fluorescent sensors for miRNA detection and imaging has increasingly attracted more attention [12, 13]. Various nanomaterials including but not limited to gold nanoparticles (AuNPs) [14, 15], quantum dots (QDs) [16], upconversion nanoparticles (UCNPs) [17, 18], and metal-organic frameworks (MOFs) [19, 20] have been effectively exploited for fluorescent sensing of miRNA. Among them, due to the inherent superiority of high porosity and good biocompatibility, Zr-MOFs possess tremendous application potential in miRNA sensing and imaging. Meanwhile, the one-step preparation of DNA-functionalized MOFs was realized by the interaction between the unsaturated metal sites of the Zr_6 MOFs and the phosphate-terminated DNA through Zr-O-P bonds, which avoids complex modifications and probe preparation processes [21].

Herein, we constructed a new ratiometric fluorescent probe based on DNA-functionalized MOFs for miRNA-21 detection and imaging in living cells with high sensitivity and specificity. Figure 1 illustrates the preparation of the probe and its application in intracellular miRNA-21 imaging. The synthesized Zr-MOFs (UiO-66-NH_2) were adopted as carriers to encapsulate Rhodamine B (RhB). DNA with a phosphate group conjugated to the UiO-66-NH_2 surface via Zr-O-P bonds to form the DNA-functionalized MOFs. Subsequently, thiazole orange (TO) was preferentially inserted into the DNA duplex to obtain a ratiometric fluorescent probe (TO-DNA-RhB@UiO-66-NH_2) based on fluorescence resonance energy transfer (FRET) between the RhB and TO. Through strand replacement reaction (SDR), the probe can specifically target miRNA-21 and recover the fluorescence of TO, hence implementing the detection and imaging of miRNA-21 in living cells. In this protocol, we describe the synthesis of the RhB@UiO-66-NH_2, the preparation of the probe TO-DNA-RhB@UiO-66-NH_2, and the detection of miRNA-21 in vitro and imaging in live cells.

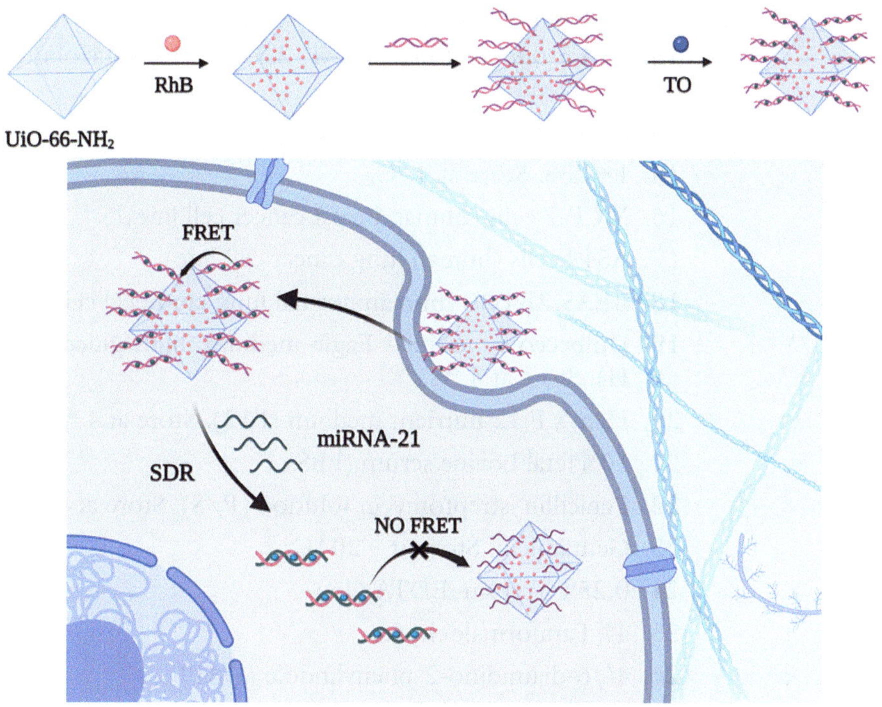

Fig. 1 Schematic illustration of the preparation of the probe TO-DNA-RhB@UiO-66-NH_2 and its application in the imaging of intracellular miRNA-21

2 Materials

All reagents are of analytical grade and without further treatment before use unless otherwise noted. Buffers and solutions are prepared and stored at room temperature. The water used in the experiment is ultrapure water (>18.2 MΩ·cm).

2.1 Reagents

1. Zirconium chloride ($ZrCl_4$). Store in the dryer.
2. 2-Aminoterephthalic acid (BDC-NH_2).
3. N, N′-dimethylformamide (DMF).
4. Acetic acid.
5. Methanol.
6. Rhodamine B (RhB). Store in dark.
7. Tris (hydroxymethyl) aminomethane (Tris).
8. Hydrochloric acid.
9. Sodium chloride.
10. Thiazole orange (TO).
11. 75% ethanol.

12. Trypan blue.
13. 3-(4,5-dimethylthiazol-2-yl)-2,5-diphenyltetrazolium bromide (MTT).
14. Dimethyl sulfoxide (DMSO).
15. 1× PBS. Store at 4 °C.
16. MCF-7 cells (human breast cancer cell line).
17. A549 cells (human lung cancer cell line).
18. BEAS-2B cells (human normal lung epithelial cell line).
19. Dulbecco's modified Eagle medium/high glucose (DMEM-H). Store at 4 °C.
20. Ham's F 12 nutrient medium (F12). Store at 4 °C.
21. 10% fetal bovine serum (FBS).
22. Penicillin–streptomycin solution (P/S). Store at −20 °C.
23. Gentamicin. Store at −20 °C.
24. 0.25% Trypsin–EDTA (1×).
25. 4% Paraformaldehyde.
26. 4′, 6-diamidino-2-phenylindole (DAPI).

2.2 Oligonucleotides DNA

1. p-DNA, 5′-TTTTTTTTTTTAGCTTATCAGA-3′, with phosphate on the 5′ terminus.
2. DNA1, 5′- TCAACATCAGTCTGATAAGCTA -3′, the miRNA-21 complementary sequence.
3. p-DNA-FAM, 5′-TTTTTTTTTTTAGCTTATCAGA-3′, with phosphate on the 5′ terminus, and FAM on the 3′ terminus.
4. Synthetic miRNA-21, 5′- UAGCUUAUCAGACUGAU GUUGA-3′.

2.3 Buffers and Reaction Solutions

1. Buffer 1: 20 mM Tris–HCl, pH 7.4, containing 100 mM NaCl. Pipet 330 μL concentrated HCl (about 12 mol/L), and dilute it to 40 mL (about 0.1 mol/L). Weigh and dissolve 242.28 mg Tris and 584.4 mg NaCl into 60 mL ultrapure water. After complete dissolving, add diluted HCl to it, and adjust pH to 7.4. Add water to a total volume of 100 mL (*see* **Note 1**).
2. Buffer 2: 20 mM Tris–HCl, pH 7.0.
3. Oligonucleotide stock solution: Oligonucleotides purchased from Sangon are in the form of dry powder. A certain amount of ultrapure water or buffer solution needs to be added to prepare a 100 μM stock solution (*see* **Note 2**).
4. RhB solution: Weigh 3.83 mg of RhB powder, and disperse it in a brown vial containing 4.0 mL of ultrapure water (*see* **Note 3**).

5. TO solution: Pipette 10 μL of TO stock solution (1 mM), and dilute to 500 μL to obtain a 20 μM TO solution (*see* **Note 4**).
6. MTT solution: 5.0 mg/mL (0.5%), dispersed in PBS. Add 1.0 mL PBS to the vial containing 100 mg MTT powder, and transfer to a 50 mL centrifuge tube after pipetting evenly. After washing the vial three times with 1.0 mL PBS, transfer to the centrifuge tube, and add PBS to make up to 20 mL, and subsequently filter it through a 0.22 μm filter to remove the bacteria in MTT solution (*see* **Note 5**).
7. Fixative solution: 4% paraformaldehyde, dispersed in PBS.
8. Staining solution: 1.0 $\mu g{\cdot}mL^{-1}$ DAPI solution, dispersed in ultrapure water. Store in a dry place at −20 °C and avoid light exposure.
9. Culture medium for MCF-7 cells and BEAS-2B cells: DMEM-H medium, supplemented with 10% (v/v) fetal bovine serum, 1% (v/v) penicillin–streptomycin and 0.1% (v/v) gentamycin sulfate.
10. Culture medium for A549 cells: F12 medium, supplemented with 10% (v/v) fetal bovine serum, 1% (v/v) penicillin–streptomycin, and 0.1% (v/v) gentamycin sulfate.

2.4 Equipments

1. Centrifuge.
2. Vacuum drying oven.
3. Cell counting plates.
4. Metal bath.
5. 6-well and 96-well cell culture plates.
6. 10-cm cell culture dishes.
7. CO_2 incubator.
8. FS 05 steady-state transient fluorescence spectrometer (Thermo Field, USA).
9. Biotech Ti-E inverted fluorescence microscope (Nikon, Japan).

3 Methods

3.1 Preparation of the RhB@UiO-66-NH_2

1. Dissolve 0.0233 g $ZrCl_4$ (0.1 mmoL) and 0.0181 g BDC-NH_2 (0.1 mmoL) in 10.21 mL DMF, and sonicate for 10 min (*see* **Note 6**).
2. Then, add 2.29 mL acetic acid to the mixture followed by sonication for another 20 min (*see* **Notes 6** and 7).
3. The mixture was then sealed in a 25 mL glass vial and maintained in a vacuum drying oven at 120 °C for 16 h.

4. After allowing the reactor to cool to room temperature naturally, centrifuge and wash the resulting yellow precipitate three times at 8000 rpm with DMF and methanol, respectively (*see* **Note 8**).
5. Soak the precipitate in methanol for 2 days, and centrifuge at 8000 rpm for 15 min every 12 h, followed by drying under the vacuum oven at 70 °C for 24 h to obtain UiO-66-NH_2. Store it at room temperature until use.
6. Weigh 2.0 mg UiO-66-NH_2, and dissolve it in a 4.0 mL RhB (2.0 mM) aqueous solution. Under dark conditions, the solution was sonicated for 2 h and then stirred at room temperature.
7. After overnight stirring, centrifuge and wash the resulting precipitate at 10000 rpm with ultrapure water until the free RhB were completely removed (*see* **Notes 9** and **10**).
8. The precipitate was dried in a vacuum drying oven at 120 °C for 12 h to obtain RhB@UiO-66-NH_2 powder. Store in a refrigerator at 4 °C for further use (*see* **Note 11**).

3.2 Preparation of TO-DNA-RhB@UiO-66-NH_2

1. Blend 10 μL p-DNA (100 μM) with 10 μL DNA1 (100 μM) in a 1.5 mL Eppendorf tube.
2. Heat the mixture at 95 °C for 5 min utilizing a mini metal bath followed by cooling down to room temperature progressively to obtain the p-DNA-DNA1 duplex (*see* **Note 12**).
3. 20 μL of the phosphate-terminated p-DNA-DNA1 hybrid (50 μM) was added to 80 μL RhB@UiO-66-NH_2 aqueous solution (2.0 mg·mL^{-1}) and placed on a shaker to incubate for 8 h in the dark. 50 μL NaCl (1.5 M) was introduced to the above mixture gradually and evenly with a time interval of 80 min to the final concentration of 0.5 M during incubation (*see* **Note 13**).
4. Centrifuge and wash the resultant solution with buffer 1 for five times to remove unattached p-DNA-DNA1.
5. The prepared DNA-RhB@UiO-66-NH_2 was resuspended in buffer 1 and stored at 4 °C for subsequent use (*see* **Note 14**).
6. Mix 100 μL DNA-RhB@UiO-66-NH_2 with TO solution (20 μM) in buffer 2 at a molar ratio of 1:4, and maintain in dark to ensure the intercalation of TO into DNA-RhB@UiO-66-NH_2 optimally (*see* **Note 15**).
7. After 5 h, centrifuge and wash the probe three times at 8000 rpm for 10 min each.
8. Finally, resuspend the resulting ratiometric probe TO-DNA-RhB@UiO-66-NH_2 in buffer 1, and store it at 4 °C.

3.3 Detection of miRNA-21 In Vitro

Since the emission spectrum of TO (energy donor) overlaps significantly (more than 30%) with the absorption spectrum of RhB (energy acceptor), FRET occurs when the insertion of TO brings TO and RhB close enough to each other, which results in a decrease in the fluorescence intensity of TO at 530 nm and an increase in the fluorescence intensity of RhB at 578 nm. In the presence of miRNA-21, it hybridizes with DNA1 through strand displacement reaction (SDR) to form a DNA/RNA complex. Hereafter, TO preferentially intercalates into the DNA/RNA complex, resulting in the spatial separation of TO and RhB, thereby hindering FRET. By measuring the fluorescence intensity ratio of TO to RhB (F_T/F_R), thus the quantitative determination of miRNA-21 was realized.

1. Dilute 100 μL of probe tenfold, and divide into ten aliquots in 1.5 mL Eppendorf tubes containing 500 μL buffer 1 each.
2. Add 100 μL of different concentrations of synthetic miRNA-21 to the pretreated probe tube.
3. After incubating for 1 h at 37 °C, transfer the reaction solution to colorimetric dishes.
4. Record the fluorescence spectrum of each sample from 520 to 650 nm on an excitation of 490 nm.

3.4 Cell Cytotoxicity Assay

1. Collect the cells in the exponential growth phase, and count cells with a cell counter (*see* **Note 16**).
2. Adjust the concentration of the cells to 5×10^4 cells/mL.
3. Seed 100 μL of A549 cell suspension (5000 cells/well) in a 96-well plate, and incubate at 37 °C in a humidified 5% CO_2 incubator for 24 h (*see* **Note 17**).
4. After cell adhesion was complete, replace the original culture medium with 100 μL of F12 culture medium containing different concentrations of the probe (1.6 μg/mL, 2.4 μg/mL, 3.2 μg/mL, 6.4 μg/mL, 9.6 μg/mL, and 12.8 μg/mL). Three parallel experiments were set for each concentration. Meanwhile, set up the blank group and negative control group (*see* **Note 18**).
5. After 24 h of incubation, add 10 μL MTT solution to each well, and continue to incubate for 4 h.
6. Then, suck out the supernatant discreetly followed by adding 100 μL DMSO to each well (*see* **Note 19**).
7. Shake the plate at a low speed for 10 min to entirely dissolve the blue-purple formazan.
8. Measure the absorbance signal of each well at 490 nm using an ELISA reader.

3.5 Cellular Uptake of the Probe

Using A549 cells as a model to investigate whether the probe could permeate into the cells or not.

1. Adjust the concentration of the A549 cells to 1×10^5 cells/mL.
2. 1 mL/well cell suspension was plated in every three wells of five 6-well plates (18 wells total) and incubated at 37 °C for 12 h.
3. Add 1.0 mL probe solution (3.2 μg/mL) to each well.
4. Incubate the five plates for different times (0 h, 2 h, 4 h, 6 h, 8 h) in the incubator, respectively.
5. After incubation, wash the cells with PBS three times (*see* **Note 20**).
6. Fix the washed cells by adding 1.0 mL fixative solution to each well.
7. After 20 min at room temperature, rinse the fixed cells three times with PBS, 5 min per wash.
8. Add 1.0 mL staining solution to each well to stain the cell nucleus, and wait for 15 min (*see* **Note 21**).
9. Wash the cells three times with PBS, 5 min per wash.
10. Record the fluorescent images of the cells under an inverted fluorescence microscope.

3.6 Imaging of miRNA-21 in Different Cell Lines

To evaluate the imaging ability of the probe for miRNA-21 in live cells and whether the probe can distinguish cancer and normal cells, three different cell lines were selected as models including MCF-7 cells (human breast cancer cell line), A549 cells (human lung cancer cell line), and BEAS-2B cells (human normal lung epithelial cell line) since studies have shown that the expression level of miRNA-21 in the three cells gradually decreased.

1. Three kinds of exponentially growing cells were collected and counted utilizing a cell counter.
2. Adjust the cell suspension concentration to 1×10^5 cells/mL using their respective culture medium.
3. 1 mL/well MCF-7, A549, and BEAS-2B cell suspensions were plated in 6-well plates. Incubate the plates in a CO_2 incubator at 37 °C.
4. After 12 h, add 1 mL probe solution (3.2 μg/mL) to each well (*see* **Note 22**).
5. Return the culture plates to the incubator and continue to incubate for another 6 h.
6. After washing the treated cells with PBS three times, fix the cells in 1 mL fixative solution for 20 min.
7. Rinse the fixed cells three times with PBS, 5 min per wash.
8. Add 1 mL staining solution to each well for 15 min.

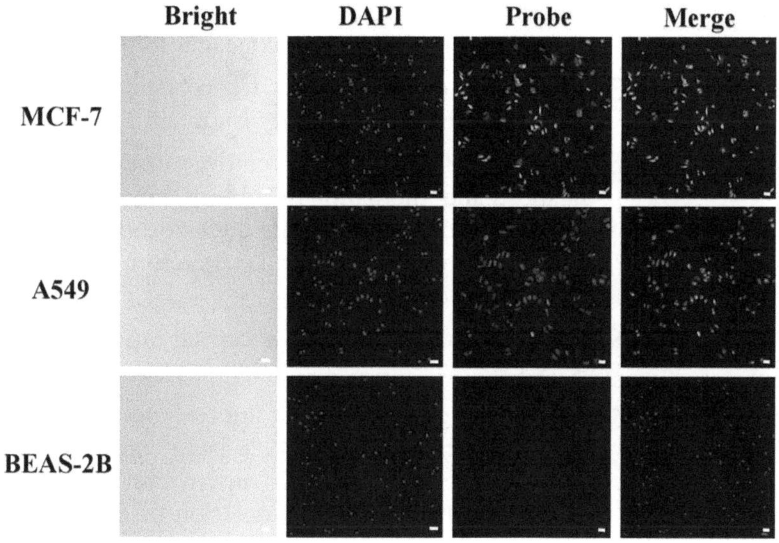

Fig. 2 Fluorescent images of MCF-7, A549, and BEAS-2B cell lines after incubation with the TO-DNA-Rh-B@UiO-66-NH_2. Scale bar: 50 μm

9. Wash the cells as **step** 7.
10. Observe all wells under an inverted fluorescence microscope, and record the fluorescent images (Fig. 2).

4 Notes

1. Since Tris–HCl buffer is susceptible to temperature and tends to absorb CO_2, store it at room temperature, and seal tightly.
2. Dilute the stock solution to the desired concentration before use, aliquot the remaining stock solution, and store it in a refrigerator at −20 °C. Avoid repeated freezing and thawing.
3. RhB solution should be prepared before use. It is photosensitive, so note to avoid light exposure during solution preparation and the subsequent operations.
4. The TO solution can be stored at −20 °C for 1 month in the dark. Store it in aliquots to avoid product failure caused by repeated freezing and thawing.
5. MTT is valid for 2 weeks when stored in the dark at 2–8 °C. Long-term storage should be at −20 °C, and avoid repeated freezing and thawing. It is best to pack it in small doses (such as 1.0 mL) and wrap it in a light-proof bag or tinfoil to prevent light from decomposing.
6. DMF and acetic acid have a pungent odor which can strongly irritate the eyes, skin, and respiratory tract. Wear a mask,

goggles, and gloves, and operate in a fume hood to reduce the harm to yourself and your colleagues.

7. Acetic acid can help adjust the particle size of UiO-66-NH_2 and improve the crystallinity to obtain a better crystal shape.
8. The role of DMF is to remove unreacted metal ions and ligands, and methanol is solvent-exchanged with DMF, to remove residual DMF in the UiO-66-NH_2 channel. Methanol has a lower boiling point and is therefore easy to be removed by subsequent drying.
9. To remove the free RhB, we carried out this process for about 3 days.
10. Collect the supernatant for fluorescence spectrometry to confirm whether the free RhB has been completely removed. The fluorescence data of the supernatant was collected from 560 to 650 nm with an excitation of 550 nm, and the max fluorescence emission at 578 nm was employed as the indicator to assess whether the free RhB is removed completely.
11. The formed RhB@UiO-66-NH_2 appears a new solid fluorescence emission peak at 578 nm compared with UiO-66-NH_2, indicating the successful introduction of RhB.
12. The cooling process should be slow, more than 2 h.
13. Both RhB@UiO-66-NH_2 and the DNA hybrid are negatively charged, so the addition of Na^+ can shield the charge and reduce the electrostatic repulsion between the DNA hybrid and RhB@UiO-66-NH_2, making it easier for the duplex to link to RhB@UiO-66-NH_2.
14. The amount of DNA connected to RhB@UiO-66-NH_2 was calculated by measuring the fluorescence intensity of the FAM-labeled DNA strand stock solution and the supernatant obtained after the functionalization.
15. Calculate the total amount of DNA according to the mass of RhB@UiO-66-NH_2 and the amount of DNA duplex contained per unit of RhB@UiO-66-NH_2, and then calculate the volume of TO that should be added according to the molar ratio.
16. Before sampling and counting, fully mix the cell suspension to ensure that the cells are in a state of uniform dispersion of single cells. After pipetting fully, immediately draw a small amount of cell suspension for counting.
17. Since the edge effect, the medium in the edge 36 wells evaporates quickly, and the drug is easy to concentrate, which has a great influence on the experiment; the edge 36 wells are filled with equivalent sterile PBS, which can also help keep the middle wells hydrated to some extent.

18. In the blank group, only medium was added, and in the negative control group, there were normal cells without probes.
19. Be careful not to destroy or remove the formazan when removing the supernatant; otherwise, it will affect the experimental results. You can try tilting the 96-well plate slightly, and then slowly suck it up with a pipette.
20. PBS can be pre-warmed in a 37 °C water bath before use.
21. After staining, the assay should be performed as soon as possible to minimize the effect of the fluorescence quenching of the dye molecules on the experiment.
22. Before this step, observe the growth state of cells under an inverted microscope during incubation; the wells with poor growth or contamination should be excluded.

Acknowledgments

This work was supported by the National Natural Science Foundation of China (No. 21904095).

References

1. Dragomir MP, Knutsen E, Calin GA (2022) Classical and noncanonical functions of miRNAs in cancers. Trends Genet 38:379–394
2. Kabekkodu SP, Shukla V, Varghese VK et al (2018) Clustered miRNAs and their role in biological functions and diseases. Biol Rev Camb Philos Soc 93:1955–1986
3. Kara G, Calin GA, Ozpolat B (2022) RNAi-based therapeutics and tumor targeted delivery in cancer. Adv Drug Deliv Rev 182:114113
4. Kunz M, Brandl M, Bhattacharya M et al (2020) Nanoparticle-complexed antimiRs for inhibiting tumor growth and metastasis in prostate carcinoma and melanoma. J Nanobiotechnol 18:173
5. Yi J, Chen T, Huo J et al (2017) Nanoscale zeolitic imidazolate framework-8 for ratiometric fluorescence imaging of MicroRNA in living cells. Anal Chem 89:12351–12359
6. Chen T, Wu Q, Cao S et al (2022) Nanodevice-based imaging of miRNA in vivo by eliminating blood interference. Chem Eng J 430:132887
7. Zhou Z, Sohn Y, Nechushtai R et al (2020) DNA Tetrahedra modules as versatile optical sensing platforms for multiplexed analysis of miRNAs, endonucleases, and aptamer-ligand complexes. ACS Nano 14:9021–9031
8. Gao J, Liu Y, Zheng B et al (2021) Light-activated and self-driven autonomous DNA nanomachine enabling fluorescence imaging of MicroRNA in living cells with exceptional precision and efficiency. ACS Appl Mater Interfaces 13:31485–31494
9. Ye M, Kong Y, Zhang C et al (2021) Near-infrared light controllable DNA Walker driven by endogenous adenosine triphosphate for in situ spatiotemporal imaging of intracellular MicroRNA. ACS Nano 15:14253–14262
10. Wang Y, Shao Z, Cheng C et al (2022) Fluorescent oligonucleotide indicators for ratiometric microRNA sensing on metal-organic frameworks. Chem Eng J 437:135296
11. Han Y, Zou R, Wang L et al (2021) An amine-functionalized metal-organic framework and triple-helix molecular beacons as a sensing platform for miRNA ratiometric detection. Talanta 228:122199
12. Lu W, Chen Y, Liu Z et al (2016) Quantitative detection of MicroRNA in one step via next generation magnetic relaxation switch sensing. ACS Nano 10:6685–6692
13. Yan T, Zhang S, Yang Y et al (2022) Biomineralization-inspired magnetic nanoflowers for sensitive miRNA detection based on exonuclease-assisted target recycling amplification. Mikrochim Acta 189:260
14. Qing Z, Luo G, Xing S et al (2020) Pt-S bond-mediated Nanoflares for high-fidelity

intracellular applications by avoiding thiol cleavage. Angew Chem Int Ed Engl 59: 14044–14048

15. Gao Y, Zhang S, Wu C et al (2021) Self-protected DNAzyme Walker with a circular bulging DNA shield for amplified imaging of miRNAs in living cells and mice. ACS Nano 15: 19211–19224
16. Martins CSM, LaGrow LP, Prior JAV (2022) Quantum dots for cancer-related miRNA monitoring. ACS Sens 7:1269–1299
17. Zhao J, Chu H, Zhao Y et al (2019) A NIR light gated DNA nanodevice for spatiotemporally controlled imaging of MicroRNA in cells and animals. J Am Chem Soc 141:7056–7062
18. Zhao T, Sun X, Chen J et al (2022) Optically programmable Plasmon enhanced fluorescence-catalytic hairpin assembly signal amplification strategy for spatiotemporally precise imaging. Anal Chem 94:5399–5405
19. Yang H, Zhao J, Dong J et al (2022) Simultaneous detection of exosomal microRNAs by nucleic acid functionalized disposable paper-based sensors. Chem Eng J 438:135594
20. Yang C, Wang K, Li Z et al (2022) A two-photon metal-organic framework nanoprobe with catalytic hairpin assembly for amplified MicroRNA imaging in living cells and tissues. Sensors Actuators B Chem 359: 131593
21. Meng H, Shi X, Chen J et al (2020) DNA amplifier-functionalized metal-organic frameworks for multiplexed detection and imaging of intracellular mRNA. ACS Sens 5:103–109

Chapter 6

3D Walking Nanomachine for Monitoring of miRNA in Living Cells

Hong-Min Meng, Xing Wang, and Zhaohui Li

Abstract

Three-dimensional (3D) DNA walking nanomachines have been widely applied in molecular diagnostics and cancer therapy. Specially, DNA dendrimer with 3D nanostructure has gained increasing attention in biomedical applications because of its high biostability, excellent programmability, and outstanding biocompatibility. In this chapter, we developed a 3D DNA walking nanomachine by assembling all nanomachine components (DNAzyme and substrate) onto a DNA dendrimer for miRNA imaging in living cells. In this design, the number of DNAzyme and substrate on the surface of the DNA dendrimer could be precisely regulated to fully implement the function of the DNA nanomachine.

Key words 3D DNA walking nanomachine, DNA dendrimer, DNAzyme, miRNA, Cell imaging

1 Introduction

DNA walking nanomachines have received intense attention and are widely used in molecular diagnostics and cancer therapy [1–3]. Among them, gold nanoparticle (AuNPs) based 3D DNA walking nanomachines that were prepared via the Au-S bond have been applied in the sensing and imaging of multitudinous bioactive species in vitro and in vivo due to their tunable optical properties, elevated cellular transfection, and improved resistance to nuclease degradation [4–6]. But, compared with one-dimensional DNA nanomachines or two-dimensional DNA nanomachines [7, 8], AuNPs-based 3D walking nanomachines lack controllability and directionality [9]. Therefore, the executive ability of each walker could not be maximumly utilized, leading to the limited theranostic applications. In addition, abundant biothiols in the complicated biological system could destroy the Au-S bond, and the stability of the nanomachine was hampered, generating false-positive signals [10, 11]. Therefore, efficient 3D DNA walking nanomachine

Shin-ichi Sato (ed.), *Live-Cell RNA Imaging: Methods and Protocols*, Methods in Molecular Biology, vol. 2875, https://doi.org/10.1007/978-1-0716-4248-1_6,

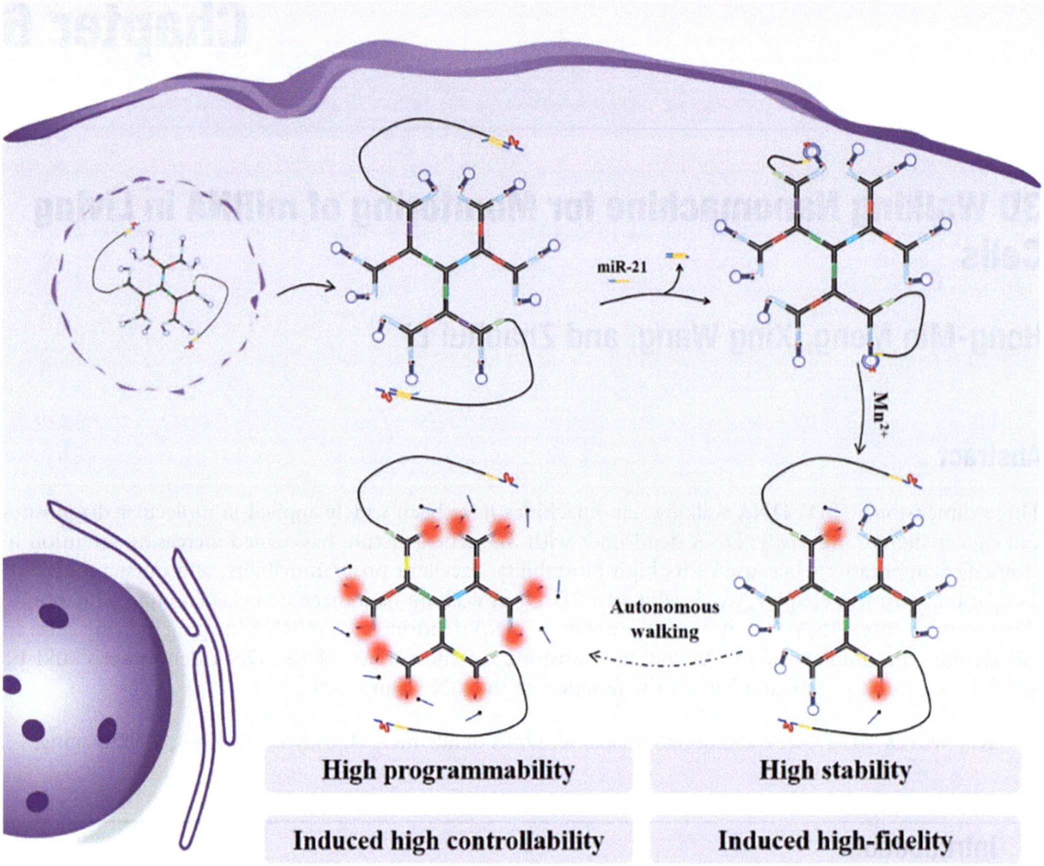

Fig. 1 Schematic illustration of the directed 3D DNA walking nanomachine for sensing miR-21 in living cells

alternatives with potentiated stability, directionality, and controllability are urgently needed.

The booming of DNA nanotechnology [12] affords an opportunity to develop fungible 3D DNA walking nanomachines. DNA dendrimer with 3D nanostructure has gained increasing attention in biomedical applications because of its high biostability, excellent programmability, and outstanding biocompatibility [13–17]. What's more, different types of functional nucleic acids (such as DNAzyme, molecule beacon, aptamer) have been combined with DNA dendrimers via the Watson-Crick base pairing for target detection, imaging, and therapy [18–23]. These intrinsic properties endow them with huge potential for constructing 3D DNA walking nanomachines for intracellular target sensing and cancer therapy, which has barely been explored.

In this chapter, we report a 3D DNA walking nanomachine by assembling all nanomachine components onto a DNA dendrimer that addresses the current limitations. The principle of this nanomachine is shown in Fig. 1. The movement of the nanomachine was initiated by a cancer biomarker (microRNA). Under the help of cofactor Mn^{2+}, the DNAzyme (Dz) could cleave the RNA substrate (S), resulting in a fluorescently labeled DNA strand away from the quencher BHQ. The free Dz subsequently moved along the track and cleaved another S, causing enhanced fluorescence recovery. In this design, the S and Dz were assembled on the dendritic DNA nanostructure controllably, and the number of S and Dz on the surface of the DNA dendrimer could be precisely regulated. Therefore, this 3D DNA walking nanomachine could fully implement the function of the nanomachine and realized the target miRNA imaging in living cells.

2 Materials

Prepare all solutions using ultrapure water (prepared by ultrapure water, to attain a sensitivity of 18.2 MΩ·cm at 25 °C) and analytical grade reagents. Prepare and store all reagents at room temperature (unless indicated otherwise). Diligently follow all waste disposal regulations when disposing waste materials.

2.1 The Preparation of 3D DNA Walking Nanomachine

1. Tris–HCl (20 mM, pH 8.3): Add about 60 mL sterile water to a glass beaker. Weigh 0.3152 g Tris–HCl, 1.7530 g NaCl, and 0.1071 g $MgCl_2$, and transfer to the glass beaker. Mix and adjust pH with HCl (*see* **Note 1**). Make up to 100 mL with water. Store at 4 °C.
2. To prepare various G_3 with different numbers of S and Dz-L, different Y_n are first prepared, the details of which are as follows: Y_0 (Y_{0a}, Y_{0b}, Y_{0c}), Y_0'(Y_{0a}, Y_{0b}, Y_{0c}'), Y_1 (Y_{1a}, Y_{1b}, Y_{1c}), Y_1'(Y_{1a}, Y_{1b}', Y_{1c}'), Y_1''(Y_{1a}', Y_{1b}', Y_{1c}'), Y_1'''(Y_{1a}', Y_{1b}', Y_{1c}), Y_2 (Y_{2a}, Y_{2b}, Y_{2c}), Y_2'(Y_{2a}', Y_{2b}, Y_{2c}'). To prepare the different Y_n (20 μM), each of the three corresponding strands (100 μM) is taken 4 μL into 8 μL Tris–HCl (20 mM, pH = 8.3) and then mixed, respectively. The temperature of the mixture is adjusted to 95 °C for 5 min and then slowly cooled to 25 °C. All of Y_n are then directly used without further purification to assemble the different generations of DNA dendrimers (G_n) (*see* **Note 2**).
3. To prepare the Dz-L (20 μM), 2 μL Dz-7 (100 μM), 2 μL locker (100 μM), and 6 μL Tris–HCl are first mixed, followed by heating to 95 °C for 5 min, and then put in ice bath for 15 min. The preparation procedure of S (20 μM) is same as the above mentioned.
4. The synthetic steps for different G_n are similar. The synthesis steps are described below using $G_{3(1:5)}$ as an example: 1 μL Y_0, 1 μL Y_1, 2 μL Y_1', and 8 μL Tris–HCl are shaken and incubated for 2 h at room temperature. Then, 2 μL Y_2 and 4 μL Y_2' are

added into the mixture and shaking incubated for 3 h at room temperature. At last, 2 μL Dz-L and 10 μL S are incubated with the mixture for 3 h to form 2/3 μM $G_{3(1:5)}$ (*see* **Note 3**).

5. 5 × TBE (45 mM Tris–base, 45 mM boric acid, 5 mM EDTA, pH 8.0): weigh 27.00 g of Tris–base, 13.75 g of boric acid, and 1.462 g of EDTA; add 400 mL sterile water. Mix and adjust pH with HCl. Make up to 100 mL with water. Store at 4 °C.
6. Agarose gel: 1% agarose.
7. 6 × loading buffer.
8. SYBR Gold.

2.2 MiRNA-21 Detection in Buffer

1. 2 mM $MnCl_2$.
2. Hitachi F-7100 fluorescence spectrometer.

2.3 Anti-interference Capability of the $G_{3(1:5)}$

1. 1 mM GSH.
2. AuNPs. 13 nM AuNPs are prepared by the classical citrate reduction route [24]. Before use, all glassware is first cleaned in aqua regia (HCl/HNO_3, 3:1) and thoroughly rinsed with ultrapure water. Next, 100 mL of 0.01% $HAuCl_4$ is heated to boiling with vigorous stirring, and 3.5 mL of trisodium citrate (1%) is added quickly. Then the solution color turned from pale yellow to colorless and finally to burgundy. Boiling is continued for an additional 10 min and then left to cool to room temperature while stirring. Then it is filtered through a 0.45 μm membrane filter. Transmission electron microscopy (TEM) images indicated the particle sizes are 13 nM ± 2 nM. The prepared AuNPs are stored for the following experiments at 4 °C.
3. AuNP-S-HP-Cy5. The HP-Cy5 are modified onto the surface of the synthesized AuNP. According to literature [25], thiol-modified substrates (10 μL 100 μM HP-Cy5) are reduced by 1 μL 15 mM Tris(2-carboxyethyl) phosphine hydrochloride (TCEP) for 1 h at room temperature. To form a hairpin structure, 20 μL PBS are added, followed by heating to 95 °C for 5 min, and then put in ice bath for 15 min. 138 μL AuNP is added and topped up to 200 μL with ultrapure water. Then the mixture is frozen in a laboratory freezer (−20 °C) for 2 h. The solution is centrifuged (13,000 rpm, 30 min) to remove the free DNA and washed three times by PBS. Finally, top up to 200 μL with 20 mM Tris–HCl. Store at 4 °C.

2.4 Cell Culture

1. HepG2 cells (the human hepatocarcinoma cell line), HEK-293 cells (human embryo kidney cells), HeLa cells (human cervical carcinoma cell line), A549 cells (human lung cancer cells), and MCF-7 cells (human breast cancer cell line).

2. DMEM (Dulbecco's Modified Eagle Medium) cell medium and RPMI-1640 (Roswell Park Memorial Institute 1640 medium) cell medium: contain 100 units/mL streptomycin, 10% fetal bovine serum, and 100 units/mL penicillin.
3. 0.1 M PBS buffer (pH 7.4).

2.5 Cell Imaging

1. 20 mM confocal dish.
2. Hoechst 33342.
3. Leica TCS SP8 confocal laser scanning fluorescence microscope.

3 Methods

Carry out all procedures at room temperature unless otherwise specified.

3.1 Characterization of 3D DNA Walking Nanomachine

1. 1% agarose gel is prepared by adding 1 g agarose per 100 mL of 1× TBE buffer and heating to dissolve. Allow the agarose to cool. Pour the gel into the cast with a comb, and allow to set for at least 30 min.
2. Mix 10 μL each of DNA samples and 2 μL 6 × loading buffer and 2 μL SYBR Gold. Load 10 μL of each sample into separate wells.
3. Plug the electrodes into the electrophoresis power supply, and set to 100 V (*see* **Note 4**). Run the gel for 30 min and check how far it has run on a trans-illuminator. Run the gel until all the bands on the gel have separated.
4. Image the gel by Molecular Imager Gel Doc XR.

3.2 The Performance of the $G_{3(1:5)}$

1. 8.33 nM varied DNA nanomachine ($G_{3(1:1)}$, $G_{3(1:2)}$, $G_{3(1:3)}$, and $G_{3(1:5)}$) are mixed with 30 nM miRNA-21 in buffer. Incubate with 2 mM $MnCl_2$ for 2 h at 37 °C.
2. Detect the fluorescence signal using Hitachi F-7100 fluorescence spectrometer under an excitation of 633 nM. Test all the measurements for three times (Fig. 2) (*see* **Note 5**).

3.3 Choosing the Best Experimental Conditions

1. Optimize the base number of Arm 1 (*see* **Note 6**).
2. Optimize the ratio between the DNAzyme and locker DNA (*see* **Note 7**).

3.4 Detect miR-21 in Buffer

1. Study the sensitivity of the 3D DNA walking nanomachine: under the optimal conditions, varied concentrations of target miR-21 (0, 0.05, 0.2, 0.4, 0.6, 1, 2, 3, 5, 8, 10, 15, 20, and 30 nM) are mixed with 8.33 nM DNA nanomachine in buffer. Incubate with 2 mM $MnCl_2$ for 2 h at 37 °C.

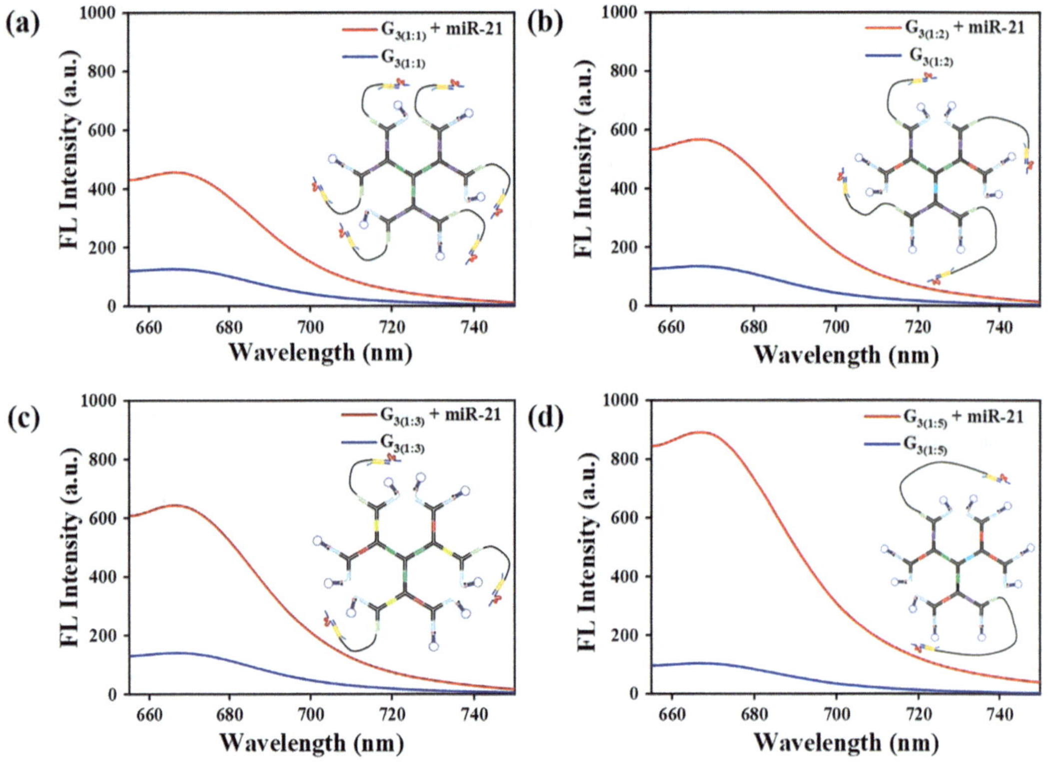

Fig. 2 Fluorescence emission spectra of 8.33 nM. (**a**) $G_{3(1:1)}$, (**b**) $G_{3(1:2)}$, (**c**) $G_{3(1:3)}$, and (**d**) $G_{3(1:5)}$ treated with 30 nM miR-21 for 2 h

2. Record the fluorescence signal using a fluorescence spectrometer under an excitation of 633 nM. Test all the measurements for three times.
3. Study the selectivity of the 3D DNA walking nanomachine: 30 nM single-base mismatched, two-base mismatched, and three-base mismatched miRNA, miR-429, miR-200b, and let-7d are mixed with 8.33 nM $G_{3(1:5)}$ in buffer. Incubate with 2 mM $MnCl_2$ for 2 h at 37 °C.
4. Record the fluorescence signal using a fluorescence spectrometer under an excitation of 633 nM. Test all the measurements for three times (Fig. 3).

3.5 Stability and Anti-Interference Capability of the $G_{3(1:5)}$

1. Study the stability of the 3D DNA walking nanomachine by gel: 10% FBS are, respectively, incubated with 1 μM $G_{3(1:5)}$ and 1 μM MB for 0, 1, 2, 3, 4, 5, 6, 7, and 8 h at 37 °C.
2. Mix 10 μL of each DNA samples and 2 μL 6 × loading buffer and 2 μL SYBR Gold. Load 10 μL of each sample into separate wells.

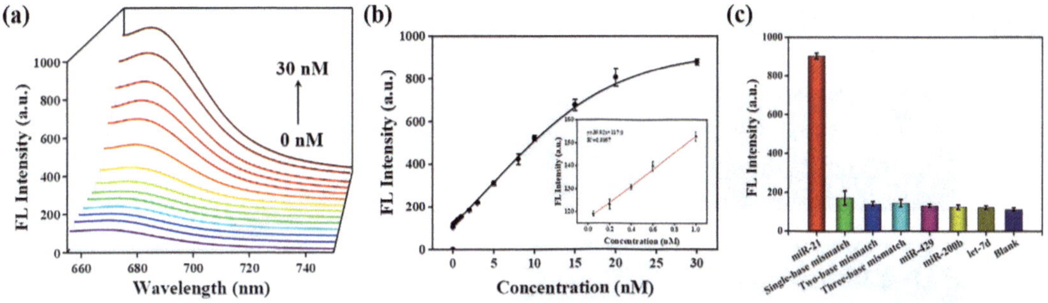

Fig. 3 (**a**) Fluorescence emission spectra of Cy5 in 8.33 nM $G_{3(1:5)}$ and (**b**) the corresponding fluorescence curve to different concentrations of miR-21 (inset shows the calibration curves of $G_{3(1:5)}$ in a concentration range of 0.05–1 nM). (**c**) Selectivity studies of $G_{3(1:5)}$ toward 30 nM different mismatched miRNA sequence. Data are mean ± SD (n = 3)

3. Plug the electrodes into the electrophoresis power supply and set to 100 V. Run the gel for 30 min, and check how far it has run on a trans-illuminator. Run the gel until all the bands on the gel have separated.
4. Image the gel by Molecular Imager Gel Doc XR.
5. Study the stability of the 3D DNA walking nanomachine by fluorescence: 10% FBS are, respectively, incubated with 10 nM $G_{3(1:5)}$ and 100 nM MB for 0, 1, 2, 3, 4, 5, 6, 7, and 8 h at 37 °C.
6. Detect the fluorescence signal using a fluorescence spectrometer under an excitation of 633 nm. Test all the measurements for three times.
7. Study the anti-interference capability of the 3D DNA walking nanomachine by fluorescence: different concentrations of GSH (0, 10, 20, 50, 100, 250, 500, 1000, 2000 μM) are, respectively, incubated with AuNP-S-HP-Cy5 and $G_{3(1:5)}$ for 12 h.
8. Detect the fluorescence signal using a fluorescence spectrometer under an excitation of 633 nM. Test all the measurements for three times.
9. 1 mM GSH are, respectively, incubated with AuNP-S-HP-Cy5 and $G_{3(1:5)}$ for 0, 2, 4, 8, and 12 h.
10. Detect the fluorescence signal using a fluorescence spectrometer under an excitation of 633 nM. Test all the measurements for three times (Fig. 4).

3.6 Detect miRNA-21 in Living Cells

1. Both normal and cancer cells are first incubated in 20 mM confocal dishes overnight to achieve 70–80% confluence.
2. Subsequently, the cells are incubated with 10 nM DNA nanomachine for 5 h. Then the cells are incubated with 1 mM Mn^{2+} for 1 h (*see* **Note 8**).

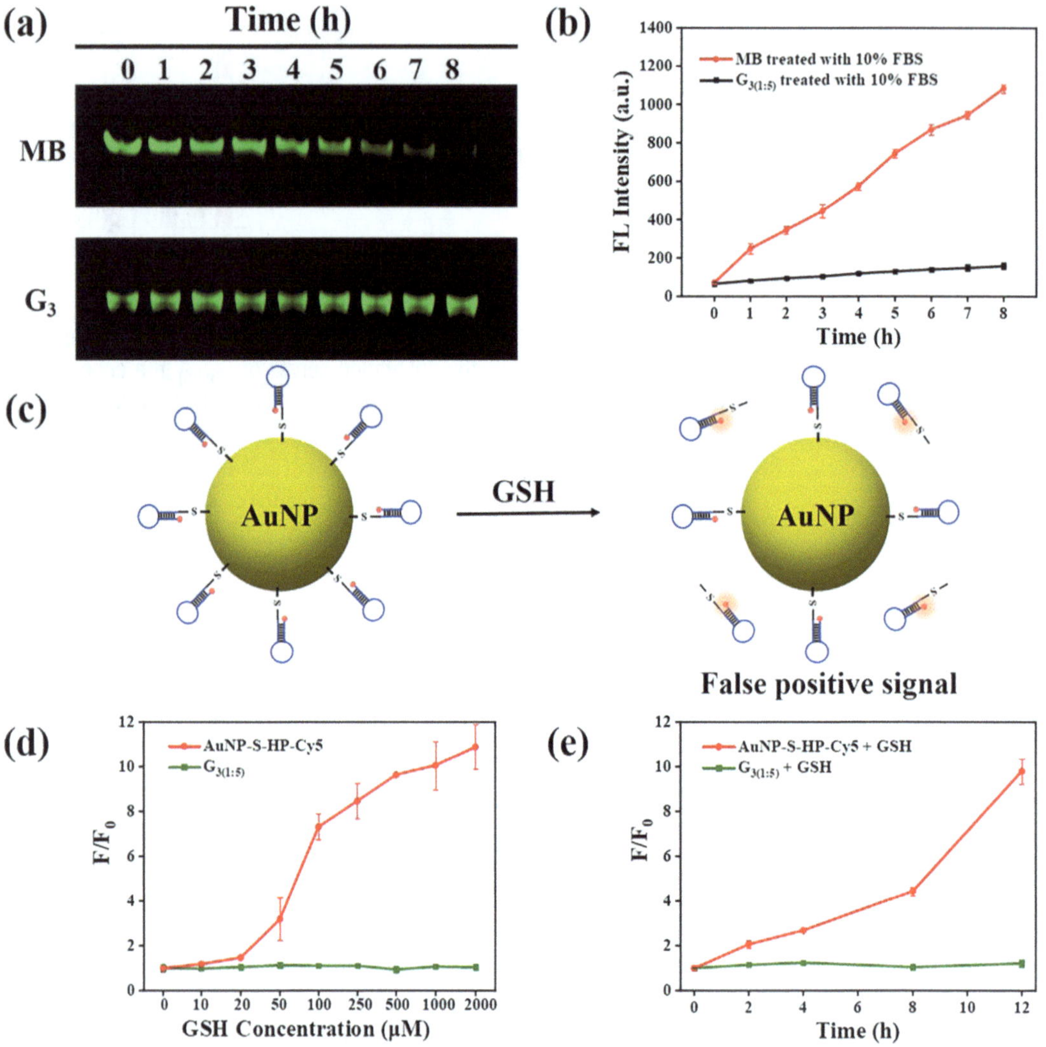

Fig. 4 (**a**) Gel electrophoresis images of free 1 μM MB and 1 μM $G_{3(1:5)}$ treated with 10% FBS for different times and (**b**) the fluorescence intensity changes of free 100 nM MB and 10 nM $G_{3(1:5)}$ treated with 10% FBS. (**c**) Schematic diagram of GSH causing false-positive signals from conventional DNA-AuNP. (**d**) Fluorescence intensity of AuNP-DNA-Cy5 and $G_{3(1:5)}$ treated with the increasing concentrations of GSH after 12 h. (**e**) Real-time monitoring of fluorescence intensity of AuNP-DNA-Cy5 and $G_{3(1:5)}$ with 1 mM GSH within 12 h

3. The cells are incubated with 20 nM Hoechst 33342 for 15 min before intracellular imaging of miRNA.
4. After washing several times with PBS (*see* **Note 9**), a confocal laser scanning fluorescence microscope is employed to image cells (Fig. 5).

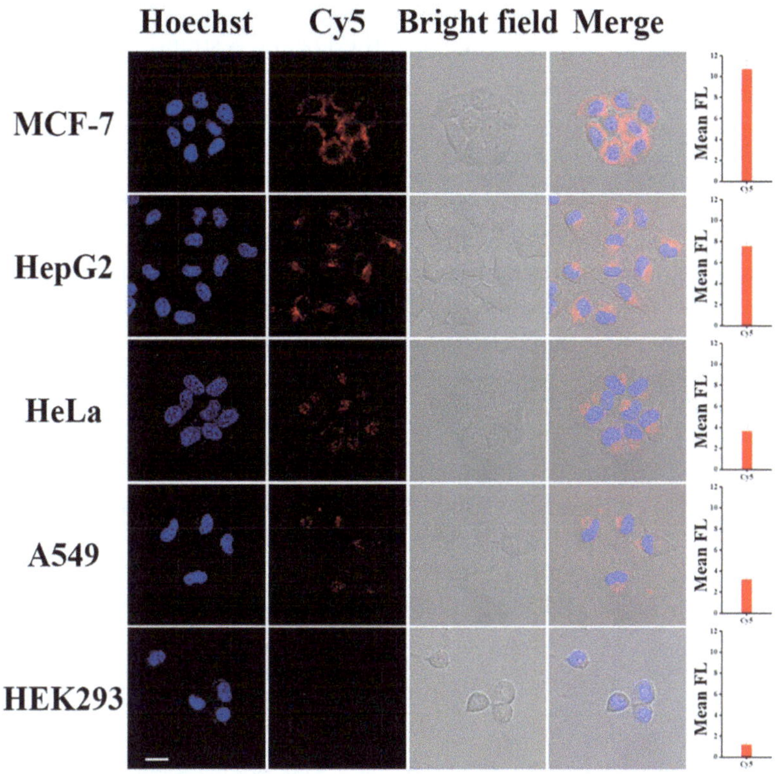

Fig. 5 Confocal images of different cell lines treated with 10 nM $G_{3(1:5)}$ for 5 h. The scale bar is 15 μm

4 Notes

1. Concentrated HCl (12 mol/L) can be used at first to narrow the gap from the starting pH to the required pH. From then on it would be better to use a series of HCl (e.g., 6 and 1 mol/L) with lower ionic strengths to avoid a sudden drop in pH below the required pH.
2. All Yn is best prepared within a week before use.
3. G_3 is best prepared within 3 days before use.
4. The voltage can be adjusted depending on the quality of the resolution and image required; after preconcentration (running at 100 V for 10 min), 80 V usually produces well-defined gels.
5. Compared with the free probe and other nanomachines, $G_{3(1:5)}$ achieved the largest fluorescence recovery at the same concentration as the target miR-21. These results suggested that the catalytic property of DNAzyme can be specifically activated by miRNA, and the sensing performance of the DNA walking nanomachine can be precisely regulated.

6. Because the base number of Arm 1 influenced the binding kinetics between the Dz and S strand.
7. Because the ratio between the DNAzyme and locker DNA influenced the binding kinetics between the Dz and locker DNA.
8. The concentration and incubation time of Mn^{2+} need to be suitable due to the potential cytotoxicity.
9. Rinsing the cells with PBS 2–3 times will help remove a bulk of the nonspecific fluorescence.

Acknowledgments

This work was supported by the Natural Science Foundation of Henan (232300421021), the National Natural Science Foundation of China (22274143, 22374136), the Program for Science Technology Innovation Teams in Universities of Henan Province (22IRTSTHN002), and Projects of the Joint Fund for Science and Technology of Henan Province (222301420007).

References

1. Hu M, Mao D, Liu X, Ren L, Zhou M, Chen X, Zhu X (2019) An all-in-one homogeneous DNA walking nanomachine and its application for intracellular analysis of miRNA. Theranostics 9:5914–5923
2. Jiang J, Zhang P, Nie Y, Peng K, Zhuo Y, Chai Y, Yuan R (2020) A well-directional three-dimensional DNA walking nanomachine that runs in an orderly manner. Chem Sci 11: 2193–2199
3. Xu Z, Liao L, Chai Y, Wang H, Yuan R (2017) Ultrasensitive electrochemiluminescence biosensor for microRNA detection by 3D DNA walking machine based target conversion and distance-controllable signal quenching and enhancing. Anal Chem 89:8282–8287
4. Aron AT, Loehr MO, Bogena J, Chang CJ (2016) An endoperoxide reactivity-based FRET probe for ratiometric fluorescence imaging of labile iron pools in living cells. J Am Chem Soc 138:14338–14346
5. Tan L, Lu J, Wang X, Liu G, Mu X, Hu K, Zhao S, Tian J (2021) A DNAzyme-driven random biped DNA walking nanomachine for sensitive detection of uracil-DNA glycosylase activity. Analyst 146:5643–5649
6. Yang Y, Huang J, Yang X, Quan K, Wang H, Ying L, Xie N, Ou M, Wang K FRET nanoflares for intracellular mrna detection: avoiding false positive signal and minimizing effect of system fluctuations. J Am Chem Soc 137(26): 8340–8343
7. Li J, Wang A, Yang X, Wang K, Huang J (2021) Orderly assembled, self-powered FRET flares for microRNA imaging in live cells. Anal Chem 93:6270–6277
8. Li Z, Li Q, Wu Y, Yuan K, Shi M, Li Y, Meng H-M, Li Z (2022) Multivalent self-assembled nano string lights for tumor-targeted delivery and accelerated biomarker imaging in living cells and in vivo. Analyst 147:811–818
9. Wen J, Chen J, Zhuang L, Zhou S (2016) Designed diblock hairpin probes for the non-enzymatic and label-free detection of nucleic acid. Biosens Bioelectron 79:656–660
10. Gao P, Liu B, Pan W, Li N, Tang B (2020) A spherical nucleic acid probe based on the Au-Se bond. Anal Chem 92:8459–8463
11. Qing Z, Luo G, Xing S, Zou Z, Lei Y, Liu J, Yang R (2020) Pt-S bond-mediated nanoflares for high-fidelity intracellular applications by avoiding thiol cleavage. Angew Chem Int Ed 59:14044–14048
12. Zhang J, Hou M, Chen G, Mao H, Chen W, Wang W, Chen J (2021) An electrochemical biosensor based on DNA "nano-bridge" for amplified detection of exosomal microRNAs. Chin Chem Lett 32:3474–3478

13. Li J, Zheng C, Cansiz S, Wu C, Xu J, Cui C, Liu Y, Hou W, Wang Y, Zhang L, Teng I, Yang H-H, Tan W (2015) Self-assembly of DNA nanohydrogels with controllable size and stimuli-responsive property for targeted gene regulation therapy. J Am Chem Soc 137:1412–1415
14. Liu R, Zhang S, Zheng T-T, Chen Y-R, Wu J-T, Wu Z-S (2020) Intracellular nonenzymatic in situ growth of three-dimensional DNA nanostructures for imaging specific biomolecules in living cells. ACS Nano 14:9572–9584
15. Meng H-M, Zhang X, Lv Y, Zhao Z, Wang N-N, Fu T, Fan H, Liang H, Qiu L, Zhu G, Tan W (2014) DNA dendrimer: an efficient nanocarrier of functional nucleic acids for intracellular molecular sensing. ACS Nano 8:6171–6181
16. Wu H, Chen T-T, Wang X-N, Ke Y, Jiang J-H (2020) RNA imaging in living mice enabled by an in vivo hybridization chain reaction circuit with a tripartite DNA probe. Chem Sci 11:62–69
17. Yue S, Song X, Song W, Bi S (2019) An enzyme-free molecular catalytic device: dynamically self-assembled DNA dendrimers for in situ imaging of microRNAs in live cells. Chem Sci 10:1651–1658
18. Geng Z, Wang L, Liu K, Liu J, Tan W (2021) Enhancing anti-PD-1 immunotherapy by nanomicelles self-assembled from multivalent aptamer drug conjugates. Angew Chem Int Ed 60:15459–15465
19. Jiang K, Wu Y, Chen J, Shi M, Meng H-M, Li Z (2021) Molecular recognition triggered aptazyme cascade for ultrasensitive detection of exosomes in clinical serum samples. Chin Chem Lett 32:1827–1830
20. Li H, Han M, Weng X, Zhang Y, Li J (2021) DNA-tetrahedral-nanostructure-based entropy-driven amplifier for high-performance photoelectrochemical biosensing. ACS Nano 15:1710–1717
21. Xue C, Zhang S, Li C, Yu X, Ouyang C, Lu Y, Wu Z-S (2019) Y-shaped backbone-rigidified triangular DNA scaffold-directed stepwise movement of a DNAzyme walker for sensitive microRNA imaging within living cells. Anal Chem 91:15678–15685
22. Yuan K, Meng H-M, Wu Y, Chen J, Xu H, Qu L, Li L, Li Z (2022) Extracellular milieu and membrane receptor dual-driven DNA nanorobot for accurate in vivo tumor imaging. CCS Chem 4:1597–1609
23. Yue R, Li Z, Wang G, Li J, Ma N (2019) Logic sensing of microRNA in living cells using DNA-programmed nanoparticle network with high signal gain. ACS Sens 4:250–256
24. Wu Y, Huang J, Yang X, Yang Y, Quan K, Xie N, Li J, Ma C, Wang K (2017) Gold nanoparticle loaded split-DNAzyme probe for amplified miRNA detection in living cells. Anal Chem 89:8377–8383
25. Yang Y, Huang J, Yang X, Quan K, Wang H, Ying L, Xie N, Ou M, Wang K (2016) Aptazyme-gold nanoparticle sensor for amplified molecular probing in living cells. Anal Chem 88:5981–5987

Chapter 7

Live-Cell Imaging of RNA G-Quadruplex with a Dual-Color Fluorescence Switch-on Probe

Takashi Sakamoto

Abstract

In-cell imaging of the G-quadruplex (G4) formation of nucleic acids remains challenging for revealing G4's functions in cells. This study describes the cell imaging method of G4 nucleic acids using our unique tripodal quinone-cyanine fluorescent dye, QCy(MeBT)$_3$, whose 600 nm and 700 nm fluorescence is enhanced independently upon the binding with double-stranded DNA and G4-DNA/RNA, respectively. The use of QCy(MeBT)$_3$ enables us to visualize cytosolic G4 RNAs in fixed and living HeLa cells with near-infrared 700 nm fluorescence.

Key words G-quadruplex, RNA, DNA, Fluorescent probe, Dual color, Quinone-cyanine dye

1 Introduction

Recent studies on the G-quadruplex (G4) structure in nucleic acids have revealed the correlation between the G4 formation of nucleic acids and various biological events [1]. G4 formation of RNAs in cytosol plays important roles in the cytosolic events in cells, such as translation [2–4] and liquid–liquid phase separation (LLPS) [5, 6]. And there is a strong correlation between these events and the prevalence of cancer [7, 8] and neurodegenerative diseases such as amyotrophic lateral sclerosis (ALS) [9, 10]. To understand G4's roles in biological events and the pathology in various diseases, the use of several fluorescent probes to visualize G4s in cells has been reported [11–15]. However, there have been few reports on fluorescence switch-on probes that can clearly distinguish between double-stranded and G4 nucleic acids. This has decreased the progress of the studies on G4 functions in cells.

Our research group recently developed tripodal quinone-cyanine fluorescent dyes that can be used as fluorescent probes for G4 imaging [16–18]. Among these fluorescent probes, QCy(MeBT)$_3$, can detect individually double-stranded DNAs

Shin-ichi Sato (ed.), *Live-Cell RNA Imaging: Methods and Protocols*, Methods in Molecular Biology, vol. 2875,
https://doi.org/10.1007/978-1-0716-4248-1_7,

(dsDNA(s)) and G4 DNA/RNAs with different wavelength fluorescence turn-on responses [16, 17]. Using QCy(MeBT)$_3$, fluorescence imaging of dsDNAs and G4 DNA/RNAs with 600 nm and 700 nm fluorescence, respectively, in fixed (Fig. 1) and living cells (Fig. 2) was successfully achieved. Herein, procedures for the

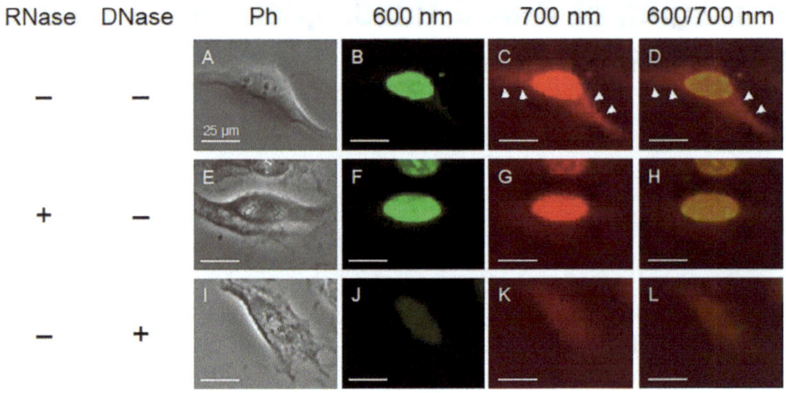

Fig. 1 Fluorescence microscopic images of fixed HeLa cells stained with QCy(MeBT)$_3$. After fixation, untreated (A–D), RNase treated (E–H), and DNase treated (I–L) cells were stained with QCy(MeBT)$_3$. Phase contrasted images (A, E, I), 600 nm fluorescence images (B, F, J), 700 nm fluorescence images (C, G, K), and merged images of 600 nm and 700 nm fluorescence (D, H, L) were collected by a fluorescence microscope BZ-X800 (KEYENCE). Scale bar: 25 μm. From the comparison between C and G or D and H, the RNase treatment caused the complete disappearance of red colored (700 nm) fluorescence in the cytosol (arrow heads in C and D), suggesting that cytosolic G4 RNAs can be imaged by QCy(MeBT)$_3$

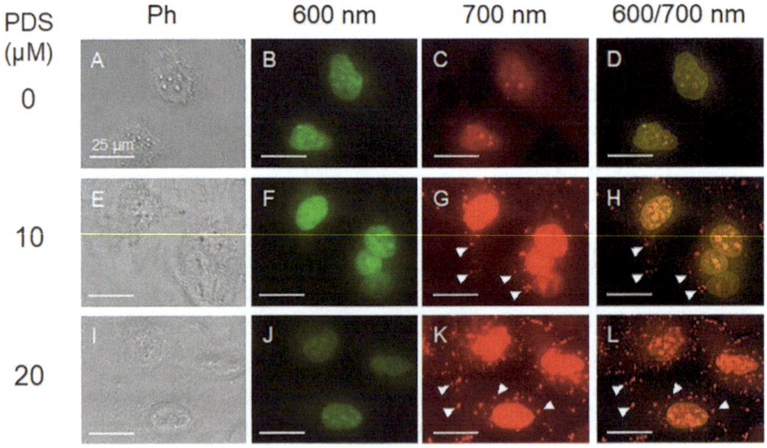

Fig. 2 Fluorescence microscopic images of living HeLa cells stained with QCy(MeBT)$_3$. Cells were treated with 0 μM (A–D), 10 μM (E–H), and 20 μM (I–L) pyridostatin (PDS). Phase contrasted images (A, E, I), 600 nm fluorescence images (B, F, J), 700 nm fluorescence images (C, G, K), and merged images of 600 nm and 700 nm fluorescence (D, H, L) were collected by a fluorescence microscope BZ-X800 (KEYENCE). Scale bar: 25 μm. From the comparison between C, G, and K or D, H, and L, the RNA foci induced by the G4 stabilization effect of PDS can be visualized by red-colored (700 nm) fluorescence in the cytosol (arrow heads in G, H, K, and L)

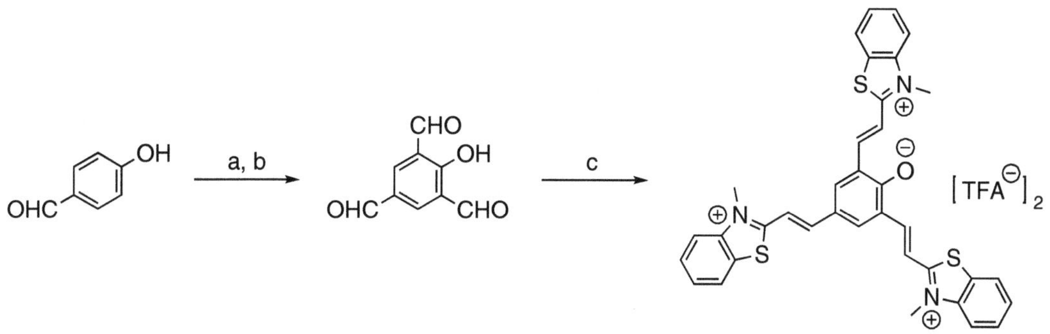

Scheme 1 Preparation of tripodal quinone-cyanine fluorescent dye, QCy(MeBT)$_3$. (**a**) Hexamethylenetetramine, trifluoroacetic acid, 130 °C; (**b**) 3 M HCl, 130 °C; (**c**) *N*-methyl-2-methylbenzothiazolium iodide, aniline, ethanol, 90 °C

synthesis of QCy(MeBT)$_3$ (Scheme 1) and for G4 RNA imaging in cells using QCy(MeBT)$_3$ are described.

2 Materials

Chemicals and solvents (organic synthesis grade) should be purchased from a general reagent supplier and used as is. Aqueous solutions should be prepared using ultrapure water (prepared by purifying deionized water, to attain a sensitivity of 18 MΩ-cm at 25 °C).

2.1 Cell Culture, Fixation, and Staining

1. Microwell plate: 24-well or 96-well.
2. CO_2 incubator.
3. Clean bench.
4. Growth medium: D-MEM (low glucose, 500 mL) supplemented with fetal bovine serum (56 mL) and penicillin–streptomycin solution (10,000 units/mL, 5.7 mL).
5. Trypsin solution: 10× Trypsin solution (Trypsin (0.5 w/v%) and EDTA· 4Na (5.3 mM) diluted with PBS.
6. Fixation solution: acetic acid (10 mL) mixed with methanol (30 mL).
7. DNase solution: Recombinant DNase I diluted in PBS (200 U/mL) (*see* **Note 1**).
8. RNase solution: RNase diluted in PBS (200 U/mL) (*see* **Note 1**).
9. Pyridostatin solution: Pyridostatin dissolved in dimethyl sulfoxide (DMSO) (3 mM) (*see* **Note 2**).
10. Staining solution: DMSO stock of QCy(MeBT)$_3$ (3 mM) and pyridostatin (3 mM) should be diluted with aqueous glucose (5%) solution or with PBS (*see* **Note 3**).

2.2 Fluorescence Imaging

1. A fluorescence microscope equipped with an objective lens (Plan Fluor 20×) and various filter blocks; excitation filter 470/40, dichroic mirror 495, and emission filter 605/70 (for imaging of 600 nm fluorescence of QCy(MeBT)$_3$); excitation filter 560/40, dichroic mirror 585, and emission filter 700/75 (for imaging of 700 nm fluorescence of QCy(MeBT)$_3$) (*see* **Notes 4** and **5**).

3 Methods

3.1 Preparation of 2-Hydroxybenzene-1,3,5-Tricarbaldehyde (See Note 6)

1. In a 200 mL two-neck flask with a magnetic stir bar and a condenser, 4-hydroxybenzaldehyde (6.1 g, 50 mmoL) should be dissolved in trifluoroacetic acid (30 mL).
2. Add hexamethylenetetramine (14.0 g, 100 mmoL).
3. N_2 substitution should be conducted.
4. The flask should be heated (130 °C), and the reaction should be refluxed under N_2 atmosphere on a hot stirrer equipped with an aluminum beads bath.
5. Cool the reaction to room temperature after 24 h refluxing.
6. Add 3 M HCl (80 mL) and reflux it again.
7. Cool the reaction to room temperature after 6 h.
8. Collect the generated precipitate by filtration through a glass filter.
9. Wash the precipitants with water and cooled ethanol, and then dry it under vacuum.
10. The title compound should be obtained as a light-yellow solid (*see* **Note** 7 and **Note 8**).

3.2 Preparation of Tripodal Quinone-Cyanine Dye

1. In a 30 mL two-neck flask with a magnetic stir bar and a condenser, 2-hydroxybenzene-1,3,5-tricarbaldehyde (9 mg, 0.05 mmoL) should be suspended in ethanol (5 mL).
2. After the addition of aniline (14 μL, 0.15 mmoL), the flask should be heated at 90 °C on a hot stirrer equipped with an aluminum beads bath.
3. Add *N*-methyl-2-methylbenzothiazolium iodide (50 mg, 0.17 mmoL) after 10 min.
4. The reaction should be monitored by reverse-phase HPLC with a 540 nm absorbance unit (*see* **Note 9**).
5. After 6–12 h, the reaction should be evaporated by a rotary evaporator.

3.3 Purification of Tripodal Quinone-Cyanine Dye

1. Dissolve the residue of the reaction in aqueous acetonitrile (30 mL, acetonitrile/water = 2:8) containing 0.1% trifluoroacetic acid.
2. Three reverse-phase column devices (Sep-Pak® Plus C18 cartridge) should be connected tandemly.
3. Wash the column devices (acetonitrile containing 0.1% trifluoroacetic acid (10 mL)) and initialized (20% acetonitrile containing 0.1% trifluoroacetic acid (10 mL)).
4. A part of the solution of the reaction residue (10 mL) should be applied to the tandemly connected column devices.
5. Elute the target fluorescent dye with aqueous acetonitrile containing 0.1% trifluoroacetic acid. The contents of acetonitrile and the volume of the eluents should be as follows: 0% acetonitrile (5 mL), 10% acetonitrile (5 mL), 20% acetonitrile (20 mL), 30% acetonitrile (30 mL), 100% acetonitrile (10 mL).
6. Items 3, 4, and 5 above should be repeated 3 times, and the 30% acetonitrile fractions were collected.
7. Lyophilize the collected fraction (90 mL) in a 500 mL round-bottom flask to give a dark-purple solid as trifluoroacetic acid salt (*see* **Note 10**).
8. Dissolve the solid in DMSO-d6 (600 μL) to analyze the structure of the compound by NMR spectroscopy (*see* **Note 11**).
9. Store the solution at −20 °C, and use it as DMSO stock for imaging G4 nucleic acids in cells.

3.4 Staining of G4 DNA/RNAs in Fixed Cells

1. Seed HeLa cells in D-MEM (10% FBS) in the wells of a 24-well culture plate (6×10^3 cells/well).
2. Culture the cells for 3 days in a humidified chamber (5% CO_2, 37 °C).
3. Wash the cells with PBS (0.5 mL).
4. After the removal of PBS, the fixation solution (25% (v/v) acetic acid in methanol) (0.5 mL) should be slowly poured into each well.
5. Incubate the cells in the fixation solution at room temperature for 10 min.
6. After removal of the fixation solution, cells should be washed with PBS (0.5 mL × 2).
7. Add DNase (0 or 200 U/mL, 0.3 mL) or RNase (0 or 200 U/mL, 0.3 mL).
8. Add staining solution ($QCy(MeBT)_3$ (6.0 μM) and pyridostatin (0 μM) in PBS, 0.3 mL).
9. Incubate the solution at room temperature for 10 min.
10. Fluorescence microscopy imaging should be performed without any washing process.

3.5 Staining of Living Cells

1. Seed HeLa cells in D-MEM (10% FBS) in the wells of a 96-well culture plate (3×10^3 cells/well), and culture them for 24 h.
2. Wash the cells with aqueous glucose solution (5% (w/v), 0.1 mL).
3. Add staining solution (QCy(MeBT)$_3$ (1.0 μM) and pyridostatin (0–20 μM) in aqueous 5% glucose (0.1 mL).
4. Wash the cells with Opti-MEM (0.1 mL) after 10 min incubation.
5. Fluorescence microscopy imaging should be performed (*see* **Note 12**).

4 Notes

1. Prepare just before use.
2. To avoid the freeze thaw cycle, DMSO stock was divided into ca. 10 μL in 0.5 mL tubes.
3. Sometimes, precipitate appears several hours after the preparation, especially in the case of the use of PBS as diluent. When the precipitate appears, you can use the staining solution after mixing of the suspension by pipetting and obtaining a homogeneous and transparent dye solution.
4. In this chapter, a fluorescence microscope BZ-X800 (KEYENCE) was used.
5. In this chapter, the fluorescence imaging was performed with a normal fluorescence microscope; however, in principle, a confocal laser scanning microscope is also available for imaging G4 nucleic acids in cells.
6. This compound is available commercially but is expensive. Thus, the synthetic method was described. The procedure used for obtaining the title compound is relatively easy.
7. Yield: 2.5 g (14 mmoL), 29%.
8. ^{1}H-NMR (400 MHz, DMSO-d6) δ 10.31 (s, 2H), 10.00 (s, 1H), 8.53 (s, 2H). Mass analysis (ESI): 177.08 (found), 177.02 (calcd. for $C_9H_5O_4^-$).
9. Conditions for HPLC were as follows: Cosmosil MS-II C18 column (2.5 μm, 3 × 70 mm) with a linear acetonitrile-H_2O (0.1% trifluoroacetic acid) gradient (1 mL/min, 20–50% acetonitrile over 10 min).
10. Yield: 8.3 mg (9.9 μmoL), 20%.
11. ^{1}H-NMR (400 MHz, DMSO-d6) δ 8.40 (d, J = 15 Hz, 2H), 8.30 (s, 2H), 8.24 (d, J = 7.6 Hz, 2H), 8.18 (d, J = 8.4 Hz,

1H), 8.12 (d, J = 8.8 Hz, 2H), 8.02 (d, J = 8.0 Hz, 1H), 8.00 (d, J = 16 Hz, 2H), 7.86 (d, J = 15 Hz, 1H), 7.74 (t, J = 8.0 Hz, 2H), 7.68 (t, J = 8.2 Hz, 1H), 7.63 (t, J = 7.4 Hz, 2H), 7.56 (t, J = 7.8 Hz, 1H), 7.51 (d, J = 16 Hz, 1H), 4.23 (s, 3H), 4.22 (s, 6H). ^{13}C-NMR (101 MHz, DMSO-d6) δ 178.4, 172.2, 170.4, 158.3, 158.0, 149.2, 146.9, 141.7, 141.7, 140.2, 129.0, 128.8, 127.7, 127.3, 126.9, 126.4, 126.2, 123.8, 123.6, 118.4, 118.4, 116.1, 115.6, 115.5, 109.9, 105.6, 35.7, 35.4. Mass analysis (ESI): 307.92 (found), 307.57 (calcd. for $C_{36}H_{29}N_3OS_3^{2+}$).

12. Time-lapse imaging of stained cells is available by the protocol using an incubation chamber that can be set up on the stage of the microscope.

Acknowledgments

This work was supported in part by a foundation from JSPS KAKENHI, Grant Number 21 K05309, and the SAMCO Science and Technology Foundation. HeLa cells were provided by RIKEN BRC, which participates in the National BioResource Project of the MEXT/AMED, Japan.

References

1. Varshney D, Spiegel J, Zyner K, Tannahill D, Balasubramanian S (2020) The regulation and functions of DNA and RNA G-quadruplexes. Nat Rev Mol Cell Biol 21(8):459–474. https://doi.org/10.1038/s41580-020-0236-x
2. Schaeffer C, Bardoni B, Mandel JL, Ehresmann B, Ehresmann C, Moine H (2001) The fragile X mental retardation protein binds specifically to its mRNA via a purine quartet motif. EMBO J 20(17):4803–4813. https://doi.org/10.1093/emboj/20.17.4803
3. Kumari S, Bugaut A, Huppert JL, Balasubramanian S (2007) An RNA G-quadruplex in the 5′ UTR of the NRAS proto-oncogene modulates translation. Nat Chem Biol 3(4):218–221. https://doi.org/10.1038/nchembio864
4. Bugaut A, Balasubramanian S (2012) 5′-UTR RNA G-quadruplexes: translation regulation and targeting. Nucleic Acids Res 40(11):4727–4741. https://doi.org/10.1093/nar/gks068
5. Asamitsu S, Shioda N (2021) Potential roles of G-quadruplex structures in RNA granules for physiological and pathological phase separation. J Biochem 169(5):527–533. https://doi.org/10.1093/jb/mvab018
6. Dumas L, Herviou P, Dassi E, Cammas A, Millevoi S (2021) G-Quadruplexes in RNA biology: recent advances and future directions. Trends Biochem Sci 46(4):270–283. https://doi.org/10.1016/j.tibs.2020.11.001
7. Wolfe AL, Singh K, Zhong Y, Drewe P, Rajasekhar VK, Sanghvi VR, Mavrakis KJ, Jiang M, Roderick JE, Van der Meulen J, Schatz JH, Rodrigo CM, Zhao C, Rondou P, de Stanchina E, Teruya-Feldstein J, Kelliher MA, Speleman F, Porco JA Jr, Pelletier J, Rätsch G, Wendel HG (2014) RNA G-quadruplexes cause eIF4A-dependent oncogene translation in cancer. Nature 513(7516):65–70. https://doi.org/10.1038/nature13485
8. Sanchez-Martin V, Lopez-Pujante C, Soriano-Rodriguez M, Garcia-Salcedo JA (2020) An updated focus on quadruplex structures as potential therapeutic targets in cancer. Int J Mol Sci 21(23):8900. https://doi.org/10.3390/ijms21238900
9. Haeusler AR, Donnelly CJ, Periz G, Simko EA, Shaw PG, Kim MS, Maragakis NJ, Troncoso

JC, Pandey A, Sattler R, Rothstein JD, Wang J (2014) C9orf72 nucleotide repeat structures initiate molecular cascades of disease. Nature 507(7491):195–200. https://doi.org/10.1038/nature13124

10. Lee YB, Chen HJ, Peres JN, Gomez-Deza J, Attig J, Stalekar M, Troakes C, Nishimura AL, Scotter EL, Vance C, Adachi Y, Sardone V, Miller JW, Smith BN, Gallo JM, Ule J, Hirth F, Rogelj B, Houart C, Shaw CE (2013) Hexanucleotide repeats in ALS/FTD form length-dependent RNA foci, sequester RNA binding proteins, and are neurotoxic. Cell Rep 5(5):1178–1186. https://doi.org/10.1016/j.celrep.2013.10.049
11. Liu LY, Liu W, Wang KN, Zhu BC, Xia XY, Ji LN, Mao ZW (2020) Quantitative detection of G-quadruplex DNA in live cells based on photon counts and complex structure discrimination. Angew Chem Int Ed 59(24):9719–9726. https://doi.org/10.1002/anie.202002422
12. Summers PA, Lewis BW, Gonzalez-Garcia J, Porreca RM, Lim AHM, Cadinu P, Martin-Pintado N, Mann DJ, Edel JB, Vannier JB, Kuimova MK, Vilar R (2021) Visualizing G-quadruplex DNA dynamics in live cells by fluorescence lifetime imaging microscopy. Nat Commun 12(1):162. https://doi.org/10.1038/s41467-020-20414-7
13. Tseng TY, Chen WW, Chu IT, Wang CL, Chang CC, Lin MC, Lou PJ, Chang TC (2018) The G-quadruplex fluorescent probe 3,6-bis(1-methyl-2-vinyl-pyridinium) carbazole diiodide as a biosensor for human cancers. Sci Rep 8(1):16082. https://doi.org/10.1038/s41598-018-34378-8
14. Di Antonio M, Ponjavic A, Radzevičius A, Ranasinghe RT, Catalano M, Zhang X, Shen J, Needham LM, Lee SF, Klenerman D, Balasubramanian S (2020) Single-molecule visualization of DNA G-quadruplex formation in live cells. Nat Chem 12(9):832–837. https://doi.org/10.1038/s41557-020-0506-4
15. Laguerre A, Hukezalie K, Winckler P, Katranji F, Chanteloup G, Pirrotta M, Perrier-Cornet JM, Wong JM, Monchaud D (2015) Visualization of RNA-quadruplexes in live cells. J Am Chem Soc 137(26):8521–8525. https://doi.org/10.1021/jacs.5b03413
16. Sakamoto T, Yu Z, Otani Y (2022) Dual-color fluorescence switch-on probe for imaging G-quadruplex and double-stranded DNA in living cells. Anal Chem 94(10):4269–4276. https://doi.org/10.1021/acs.analchem.1c04804
17. Sakamoto T, Yu Z (2022) Tripodal quinone-cyanine fluorescent dye as a tool for analyzing structural dynamics of DNA guanine-quadruplex. Chem Lett 51(12):1139–1142. https://doi.org/10.1246/cl.220420
18. Muraoka Y, Muramoto J, Yasuhara Y, Kawatake M, Sakamoto T (2023) Near-infrared fluorescent switch-on probe for guanine-quadruplex imaging with extremely large Stokes shift. Anal Chem 95(47):17162–17165. https://doi.org/10.1021/acs.analchem.3c04318

Chapter 8

Visualization of Endogenous RNA G-Quadruplex in Individual Genes in Single Cells

Yicong Dai, Xucong Teng, Hongwei Hou, and Jinghong Li

Abstract

Single-cell RNA imaging enables the detection of endogenous RNA and the analysis of its abundance, localization, and single-cell heterogeneity, which is powerful in the study of RNA biology. G-quadruplex (G4) is a kind of nucleic secondary structure and plays important roles in various biological processes. To reveal the regulating mechanism of monogenic G4s in a specific biological pathway, it is of great important to visualize endogenous RNA G4 in individual genes. Here, we describe Module Assembled Multifunctional Probes Assay (MAMPA) for single-cell visualization of monogenic RNA G4. This method works well in various cell lines and various RNA G4 analytes and has a low detection limit at ten copies per cell.

Key words G-quadruplex, RNA secondary structure, Single-cell imaging, Monogenic G-quadruplex detection, Nucleic acid templated bio-orthogonal reaction, Fluorescent probe, Rolling circle amplification

1 Introduction

Single-cell RNA imaging utilizes nucleic acid probes and fluorescent microscopy to visualize intracellular RNA with high specificity and sensitivity [1]. This technique preserves the physiological state and microenvironment of cells and has promoted the understanding of the nature of RNA, such as intracellular abundance, localization, and single-cell heterogeneity. However, most of the single-cell RNA imaging methods can only detect which gene the RNA belongs to, but cannot provide the secondary structure of RNA [2], such as G-quadruplex (G4). G4 is a sort of nucleic acid secondary structure, which can form both in DNA and RNA through the stacking of Hoogsteen hydrogen-bonded G-tetrads (Fig. 1a). In human genome and transcriptome, there are more than one million G4s, and they play important roles in many biological processes, including transcription, translation, RNA splicing, and so on [4, 5]. In recent years, G4 has been increasingly reported to

Shin-ichi Sato (ed.), *Live-Cell RNA Imaging: Methods and Protocols*, Methods in Molecular Biology, vol. 2875, https://doi.org/10.1007/978-1-0716-4248-1_8,

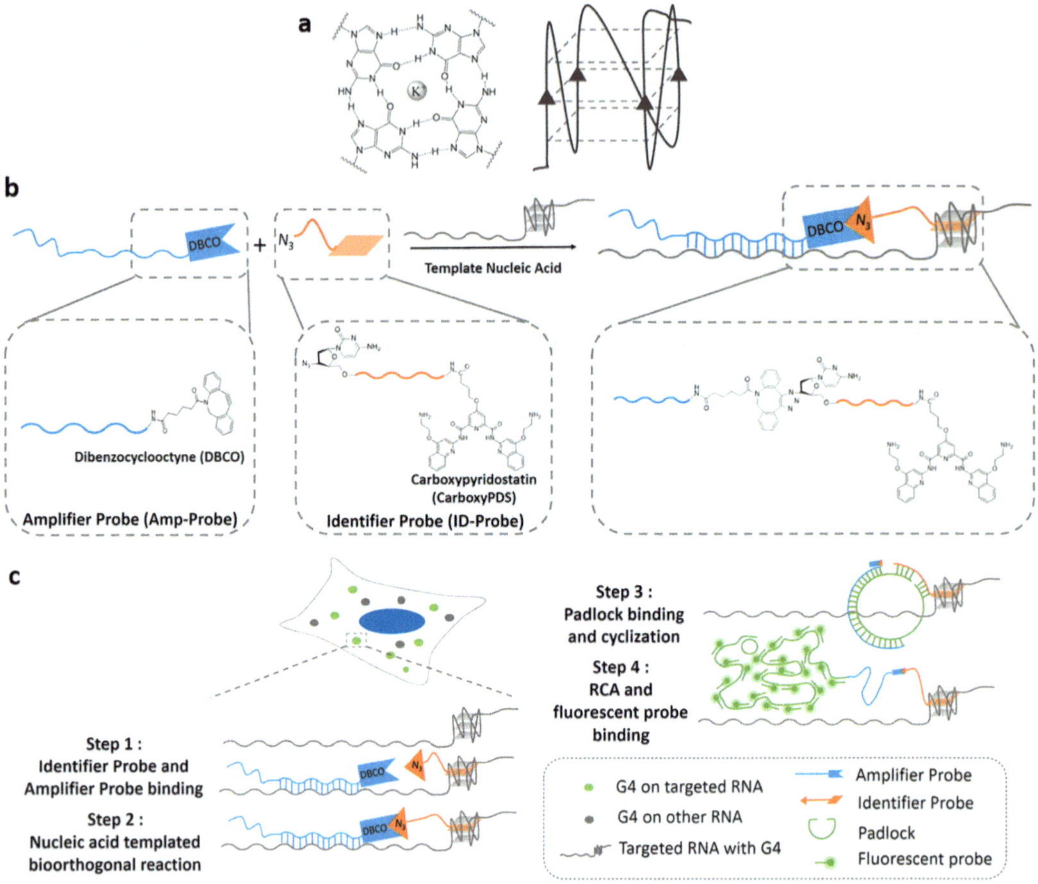

Fig. 1 (**a**) Schematic illustration of a G-tetrad (left) and a G4 structure (right). (**b**) Schematic illustration of nucleic acid templated click reaction. Amp-Probe and ID-Probe are connected through nucleic acid template DBCO-N_3 click reaction after both probes recognize the targeted RNA. (**c**) Experimental strategy of MAMPA for imaging G4s in an individual gene. (Reprinted with permission from Ref. [3]. Copyright 2021 Wiley-VCH GmbH)

become potential therapeutic target of many diseases [6–9], although the detail mechanisms are still unclear. To unravel the biofunction of a single G4-containing gene, it is of great important to visualize endogenous RNA G4 in individual genes.

Here, we introduce Module Assembled Multifunctional Probes Assay (MAMPA), the first method for single-cell visualization of monogenic RNA G4 [3]. MAMPA separately detects the G4 structure and the specific RNA sequence with an ID-Probe and an Amp-Probe (Fig. 1b). ID-Probe contains a 3′-azide modified poly-adenylic acid backbone and a CarboxyPDS unit. CarboxyPDS is a G4 ligand with low Kd (100 nM) and high specificity to RNA G4 over DNA G4 [10, 11]. Amp-probe is a 5′- dibenzocyclooctyne (5′-DBCO) modified nucleic acid probe that hybridizes to the nearby sequence of targeted RNA. Upon binding to the two target parts of the analyte RNA, ID-Probe and Amp-Probe chemically ligate together through nucleic acid templated bio-orthogonal reaction [12]. The assembled probe is recognized by a modified rolling circle amplification (RCA) process [13, 14]. Concretely, a padlock probe recognizes ID-Probe and is cyclized to serve as the circular template of RCA. Meanwhile, the free 3′-end of Amp-Probe serves as the primer of RCA. Finally, fluorophore-labeled probes bind with the RCA product to generate a fluorescent spot of a G4 in an individual RNA (Fig. 1c).

MAMPA works well in both high-abundance and low-abundance genes and shows a high specificity and sensitivity. The detection limit is as low as ten copies per cell. It has been verified in human HeLa, MCF-7, BT-474, SK-BR-3, MDA-MB-231, MCF 10A, MDA-kb2, and mouse ESC cell lines and would be applicable in all mammalian cell lines.

2 Materials

2.1 DNA Oligo Probes

1. A specific Amp-Probe and a specific padlock probe need to be designed according to the sequence of each target RNA. The Amp-Probe has the backbone of R_{20}GCTAGCTAGC. The padlock probe has the backbone of TTTTTTCCCAAC-TATACAACATACTACCTCACTT GCTAGCTAGCN_{20}TTT TTTTT. Find the G4-containing sequence nearby the target G4 motif (*see* **Note 1**). Regard the 5′ first G of the target G4 motif as 0 nt, then take the sequence from −27 nt to −8 nt as N_{20}, and fill it into the backbone sequence of the padlock probe. Derive the reverse sequence of N_{20} as R_{20}, and fill it into the backbone sequence of Amp-Probe. The example of designing the probes for human BCL2 mRNA G4 is shown in Fig. 2. Remind that Amp-Probe must be 5′-DBCO modified, and the padlock probe must be 5′-phosphorylated. The fluorescent probe, 5′-AlexaFluor488 modified AACTATACAA-CATACTACCTCA, is universal for all genes (*see* **Note 2**).
2. Purchase the designed Amp-Probe, padlock probe, and fluorescent probe from an oligo synthesis company. These probes must be HPLC purified.

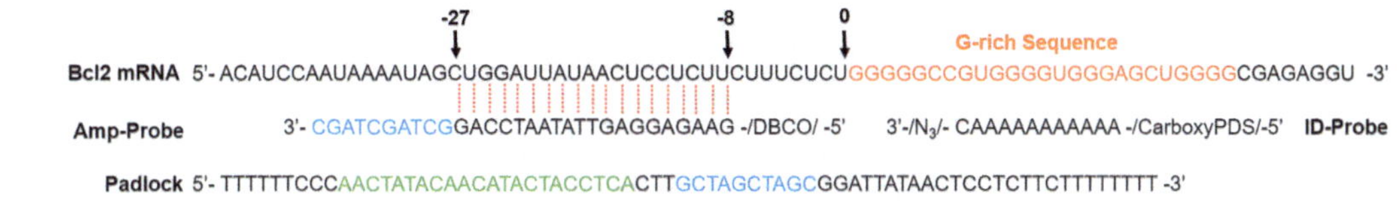

Fig. 2 The design principle of the DNA oligo probes of human BCL2 mRNA G4

3. Synthesize the ID-probe according to the published protocol [3].
4. Dissolve all the probes with 1X TE buffer (pH = 8.0) to a concentration of 10 μM. Aliquot and store them at −80 °C.

2.2 MAMPA Imaging

1. DEPC-treated 1XPBS: Add 1 g diethypyrocarbonate (DEPC) to 1 L 1XPBS in a 1.5 L glass bottle, and shake the bottle to mix it well. Incubate the solution at room temperature for 24 h. Then autoclave the solution for 25 min.
2. 4% (w/v) paraformaldehyde: Dissolve 4 g paraformaldehyde in 100 mL DEPC-treated 1X PBS (*see* **Note 3**).
3. 0.5% v/v Triton-X100: Add 50 μL Triton-X100 to 10 mL DEPC-treated 1XPBS.
4. Probe-binding solution: 2X saline-sodium citrate buffer (SSC), 1 μM Amp-Probe, 1 μM ID-Probe, 1 mg/mL yeast tRNA, 5 mM DTT (dithiothreitol), 1 U/μL RNase inhibitor (*see* **Notes 4** and **5**).
5. Padlock-binding solution: 2XSSC, 1 μM padlock probe, 1 mg/mL yeast tRNA, 5 mM DTT, 1 U/μL RNase inhibitor.
6. Padlock-cyclization solution: 1X ligase reaction buffer, 1 U/μL T4 DNA ligase, 1 U/μL RNase inhibitor (*see* **Note 6**).
7. RCA solution: 1×phi29 DNA polymerase reaction buffer, 0.4 U/μL phi29 DNA polymerase, 3 mM dNTPs, 1 U/μL RNase inhibitor.
8. Fluorescent probe-binding solution: 0.1 μM fluorescent probe, 2×SSC, 0.1 v/v formamide, 10 ng/μL salmon sperm DNA (*see* **Note** 7). Prepare it in the dark and store it away from light.

3 Methods

3.1 Cell Preparation

Culture cells in the wells of a 384-well glass-bottom plate 24 h before the MAMPA experiment. Keep the cells at about 70% confluence when conducting the experiment (*see* **Notes 8** and **9**).

3.2 MAMPA Imaging

The amount of solutions in each step is for a single well of cells.

1. Fix the cells in 50 μL 4% (w/v) paraformaldehyde for 15 min at room temperature (*see* **Note 10**).
2. Permeabilize the cells for 5 min with 50 μL 0.5% v/v Triton-X100. Then wash the fixed cells with 50 μL DEPC-treated 1X PBS for three times (*see* **Notes 11** and **12**).

3. Add 20 μL probe-binding solution and incubate the mixture at 37 °C for 2 h. Then wash the fixed cells with 50 μL DEPC-treated 1X PBS for three times.
4. Add 20 μL padlock-binding solution, and incubate the mixture at 37 °C overnight. Then wash the fixed cells with 50 μL DEPC-treated 1X PBS for three times.
5. Add 10 μL padlock-cyclization solution, and incubate the mixture at 37 °C for 2 h. Then wash the fixed cells with 50 μL DEPC-treated 1X PBS for three times.
6. Add 10 μL RCA solution and incubate the mixture at 37 °C for 2 h. Then wash the fixed cells with 50 μL DEPC-treated 1X PBS for three times.
7. Add 20 μL fluorescent probe-binding solution, and incubate the mixture at 37 °C for 0.5 h. Then wash the fixed cells with 50 μL DEPC-treated 1X PBS for three times. Conduct this procedure in the dark.
8. Mount the sample with Fluoromount-G containing 4′,6-diamidino-2-phenylindole (DAPI). Then the sample is ready for imaging (*see* **Note 13**).

3.3 Imaging and Data Analysis

1. Set a 63X oil-immersion objective in an inverted confocal microscope. Set the image size to 164 μm × 164 μm and the single pixel size to 160 nm × 160 nm.
2. Collect the images through layering with the step size of 0.2 μm to ensure that all RCA particles in cells can be collected. Collect DAPI channel (nucleus), FITC channel (RCA particles), and bright field channel (cell morphology).
3. Use the software of the confocal microscope to stack the layered images with the method of maximum intensity project (MIP).
4. Set the threshold value using the software ImageJ to distinguish the RCA particles and the nuclei of cells from the background signal. Set the RCA particles as green channel and the nuclei of cells as blue channel. Merge the RCA particles and the nuclei of cells to a single image. Manually label the outline of cells according to the bright field images. A representative result of MAMPA is shown in Fig. 3.
5. Count the numbers of RCA particles in a single cell ($n = 100$ cells in each experimental condition) (Fig. 3b).

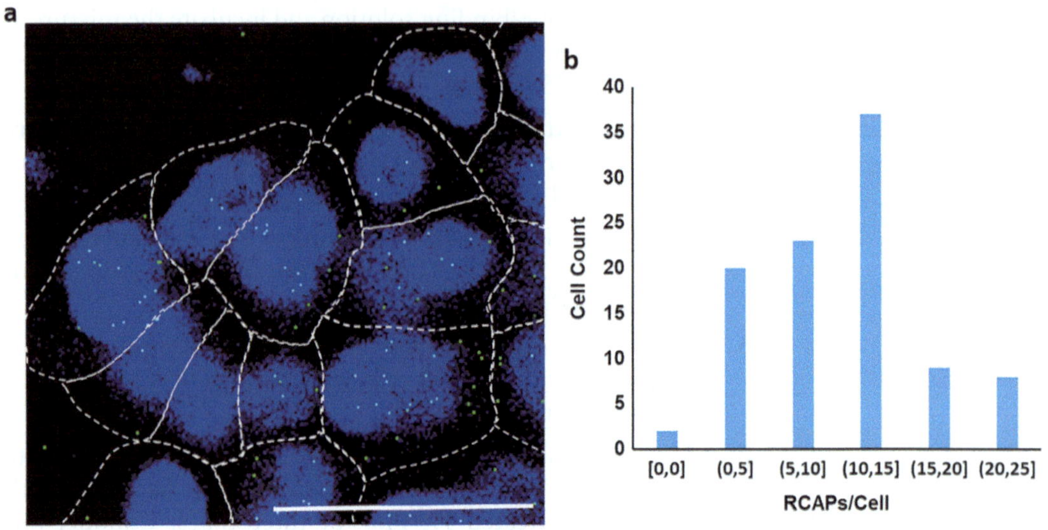

Fig. 3 Representative result of MAMPA. (**a**) Image of G4 on human BCL2 mRNA in HeLa cells. The cell nuclei are shown in blue (DAPI), and the RCA particles appear as green spots (AlexaFluor488). Cells are outlined by white lines. Scale bar = 25 μm. (**b**) Frequency histograms of RCA particles per cell. (Reprinted with permission from Ref. [3]. Copyright 2021 Wiley-VCH GmbH)

4 Notes

1. The sequence of the target RNA can be found from NCBI (https://www.ncbi.nlm.nih.gov/). The G4 motif of the target RNA can be analyzed by QGRS Mapper (https://bioinformatics.ramapo.edu/QGRS/index.php).
2. The fluorophore of the fluorescent probe can be changed according to the specific experimental condition. For example, when MitoTracker Green is used to image the mitochondria at the same time, the fluorophore of the fluorescent probe can be changed to Cy5.
3. Avoid the degradation of intracellular RNA caused by RNase in the preparation of the solutions and the MAMPA procedures. Prepare all solutions using DNase-free and RNase-free buffers and reagents. Use RNase-free tubes and pipette tips, and conduct all the operations in a clean bench.
4. DTT is easily oxidized. Aliquot and store it at −80 °C.
5. Yeast tRNA is easily degraded. Aliquot and store it at −80 °C. Do not refreeze it.
6. It is recommended to purchase the enzymes used in this method from Thermo Scientific. The enzymes from different companies may have different definition of its activity unit.

Thus, if the enzymes from other company are used, test the best amount of the enzymes in each step.

7. The amount of formamide and salmon sperm DNA can be adjusted. If the background signal is too high, increase their dosage and vice versa.
8. For poorly adherent cells, treat the 384-well plate with 0.1% gelatin at 37 °C for 0.5 h before culture the cells.
9. Wells in a 384-well plate are not reusable, but unused wells can still be used. Sterilize the 384-well plate by UV irradiation before culturing the cells.
10. The time of fixation and the time of permeabilization need to be adjusted according to the cell type. It is recommended to fix the cells for 10–20 min and to permeabilize the cells for 3–10 min.
11. The well of 384-well plate is very small. When washing the well, use 10 μL pipette tips and operate carefully. Suck and add the liquid from one corner of the well each time to avoid the loss of cells.
12. In each washing step, the cells can also be washed with DEPC-treated 1X PBS containing 0.1% Tween20. The addition of Tween20 may increase the signal-to-noise ratio.
13. The mounted sample can be stored at room temperature away from light for about 12 h. Image it as soon as possible.

Acknowledgments

This work was financially supported by Beijing Life Science Academy Initiative Scientific Research Program (No. 2023000CA0050, No. 2023100CC0240, and No. 2023100CC0250), China Postdoctoral Science Foundation (No. 2023M731972 and No. BX20230179), and the New Cornerstone Science Foundation.

References

1. Pichon X, Lagha M, Mueller F, Bertrand E (2018) A growing toolbox to image gene expression in single cells: sensitive approaches for demanding challenges. Mol Cell 71:468–480
2. Teng X, Dai Y, Li J (2022) Module assembly strategy for single-cell nucleic acid imaging at the sub-molecule level. Chem Eur J 28: e202104628
3. Dai Y, Teng X, Li J (2022) Single-cell visualization of monogenic RNA G-quadruplex and occupied G-quadruplex ratio through a module-assembled multifunctional probes assay (MAMPA). Angew Chem Int Ed 61: e202111132
4. Varshney D, Spiegel J, Zyner K, Tannahill D, Balasubramanian S (2020) The regulation and functions of DNA and RNA G-quadruplexes. Nat Rev Mol Cell Biol 21:459–474
5. Teng X, Dai Y, Li J (2022) Methodological advances of bioanalysis and biochemical

targeting of intracellular G-quadruplexes. Exploration 2:20210214
6. Neidle S (2017) Quadruplex nucleic acids as targets for anticancer therapeutics. Nat Rev Chem 1:0041
7. Collie GW, Parkinson GN (2011) The application of DNA and RNA G-quadruplexes to therapeutic medicines. Chem Soc Rev 40: 5867–5892
8. Balasubramanian S, Hurley LH, Neidle S (2011) Targeting G-quadruplexes in gene promoters: a novel anticancer strategy? Nat Rev Drug Discov 10:261–275
9. Kharel P, Balaratnam S, Beals N, Basu S (2020) The role of RNA G-quadruplexes in human diseases and therapeutic strategies. Wiley Interdiscip Rev RNA 11:e1568
10. Di Antonio M, Biffi G, Mariani A, Raiber EA, Rodriguez R, Balasubramanian S (2012) Selective RNA versus DNA G-quadruplex targeting by in situ click chemistry. Angew Chem Int Ed 51:11073–11078
11. Rocca R, Talarico C, Moraca F, Costa G, Romeo I, Ortuso F, Alcaro S, Artese A (2017) Molecular recognition of a carboxy pyridostatin toward G-quadruplex structures: why does it prefer RNA? Chem Bio Drug Design 90:919–925
12. Kukwikila M, Gale N, El-Sagheer AH, Brown T, Tavassoli A (2017) Assembly of a biocompatible triazole-linked gene by one-pot click-DNA ligation. Nat Chem 9:1089–1098
13. Deng R, Zhang K, Wang L, Ren X, Sun Y, Li J (2018) DNA-sequence-encoded rolling circle amplicon for single-cell RNA imaging. Chem 4:1373–1386
14. Gao H, Zhang K, Teng X, Li J (2019) Rolling circle amplification for single cell analysis and in situ sequencing. Trac-Trend Anal Chem 121: 115700

Chapter 9

Imaging and Quantifying Ribosomal Frameshifting Dynamics with Single-RNA Precision in Live Cells

Kenneth R. Lyon, Tatsuya Morisaki, and Timothy J. Stasevich

Abstract

Recent advances in fluorescence microscopy have now made it possible to measure the translation dynamics of individual RNA in living cells and in multiple colors. Here we describe a protocol that exploits these recent advances to simultaneously image the translation of two open reading frames encoded on a single reporter RNA yet frameshifted with respect to each other. This enables precise measurements of frameshifting dynamics and efficiency from specific frameshift stimulatory sequences, all with single-RNA precision.

Key words Single-molecule imaging, Frameshifting, Translation, Gene regulation, Viruses

1 Introduction

Technological advances over the past 8 years have made it possible to image the real-time translation of single RNAs in vivo [1–5]. To achieve this, a repeat epitope tag is used in place of a more traditional tag like GFP (green fluorescent protein). Repeat epitope tags consist of an epitope (such as HA or FLAG) repeated multiple times within a structure [6, 7]. The epitopes are bound and lit up cotranslationally by complementary fluorescent antibody-based probes coexpressed in cells. This amplifies fluorescence at translation sites since multiple probes can bind a single repeat epitope tag and multiple tags can be translated simultaneously from a single reporter mRNA as a polysome.

By using a combination of complementary repeat epitope tags/probes, it is now possible to image translation in multiple colors at the same time [8–10]. With this strategy, the translation dynamics of multiple unique reporters can be tracked in separate colors for multiplexing. Alternatively, translation from a single polycistronic reporter can be tracked in multiple colors. These types of studies have provided a fresh perspective on mRNA translation and have

Shin-ichi Sato (ed.), *Live-Cell RNA Imaging: Methods and Protocols*, Methods in Molecular Biology, vol. 2875,
https://doi.org/10.1007/978-1-0716-4248-1_9,

highlighted the surprising heterogeneity in translation dynamics that exists between a genetically identical pool of mRNA. Recent applications have explored a wide range of gene regulatory phenomena, including translational bursts and pauses [2, 3], local translation [11], ribosomal traffic jams [12], miRNA-mediated translational control [13], translation during stress [14, 15], start codon selection [10], and internal ribosomal entry site usage [16].

In the spirit of these multicolor applications, we recently developed a novel multi-frame reporter that makes it possible to image and quantify in living cells translational frameshifting dynamics with single-RNA resolution [17]. Programmed ribosomal frameshifts (PRFs) are induced by unique RNA structures that cause elongating ribosomes to shift their translational reading frames, leading to the production of different downstream protein products [18]. PRFs are commonly used by viruses to minimize their genomes and to streamline viral biogenesis by precisely controlling the ratio of frameshifted to non-frameshifted viral protein production. Many viruses that have epidemic and pandemic potential such as HIV [19], influenza [20], and SARS-CoV-2 [21–23] employ PRFs.

To capture PRF dynamics live and with single-RNA resolution, we created a multi-frame reporter construct that encodes a frameshift stimulatory sequence (FSS) followed by alternating repeat epitopes in the 0 and − 1 (or + 1) frames (Fig. 1a). We also inserted a downstream reporter sequence that codes for a functional protein of roughly the same size in both frames. With this setup, when frameshifting occurs, translation appears in a different color than it normally does, even though the timing and intensity of the translation signal remain similar (Fig. 1b).

Traditionally, frameshifting has been quantified using dual bioluminescent reporters [24]. These assays have uncovered trans-acting factors that influence PRFs and have also helped quantify how much end-point protein product is produced from the frameshifting and non-frameshifting frames. Although these types of bulk assays can characterize many important aspects of PRFs, they make it difficult to assess the real-time regulation and heterogeneity of frameshifting. This was the original motivation for developing our multi-frame tag. We figured the ability to image frameshifting at individual RNA would allow us to see if frameshifting dynamics differ from one RNA to another. Surprisingly, we found frameshifting can be highly heterogeneous, with most RNAs translating normally and only a small subset frameshifting, but with very high probability [17].

Herein we describe how to use our multi-frame tag to enable single molecule experiments that can assess the heterogeneity of frameshifting from one RNA to another. In particular, we describe the plasmids needed, tissue culture and transfection protocols, and the imaging procedure. It is our hope that this protocol will help others uncover the heterogeneity in frameshifting dynamics at a variety of PRFs in a wide range of biological systems.

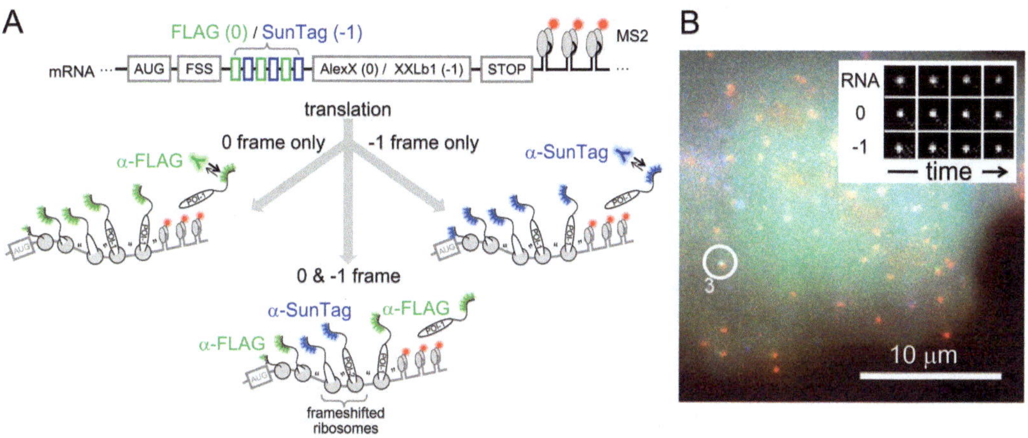

Fig. 1 Multi-frame reporter to image single-mRNA frameshifting in live cells. (**a**) Schematic of the multi-frame reporter. The reporter RNA is tagged with 24 MS2 stem loops in the 3′ untranslated region (red, MS2). After the AUG start codon, there is a frameshift stimulatory sequence (FSS), followed by 10 FLAG epitopes in the 0 frame alternating between 10 SunTag epitopes in the −1 frame. Following these repeat epitopes is a reporter protein of interest (POI) that codes for the AlexX protein in the 0 frame (626 aa; POI-1) and the XXLb1 protein in the −1 frame (689 aa; POI-2). If ribosomes (gray circle in cartoon) translate the reporter without frameshifting, only 0 frame FLAG epitopes are produced, leading to a polysome foci that is labeled with MS2 coat proteins (red) and α-FLAG antibody fragments (green). Alternatively, if ribosomes all frameshift during translation, only −1-frame SunTag epitopes are produced, leading to mRNA foci labeled with MS2 coat proteins (red) along with α-SunTag antibody fragments (blue). A final possibility is that some ribosomes frameshift, while others do not, leading to a polysome that is labeled in all three colors. (**b**) A portion of a cell expressing the multi-frame reporter, along with MS2 coat proteins to label reporter mRNA (red), α-FLAG antibody fragments conjugated with Cy3 to label 0-frame AlexX reporter protein (green), and α-SunTag antibody fragments tagged with GFP to label reporter protein XXLb1 (blue). Multiple mRNA spots are present in the cell. The circled spot (white circle labeled "3" for the three signals detected there) shows a frameshifting site with all three signals present, including mRNA (MS2, red), 0-frame reporter protein AlexX (labeled by α-FLAG antibody fragments, green), and − 1-frame reporter protein XXLb1 (labeled by α-SunTag antibody fragments, blue). The frameshifting site can be tracked through time in all three colors (upper-right inset). Scale bar, 10 μm

2 Materials

2.1 Sample Preparation

Aside from the specific reporter plasmids listed below, this is similar to another protocol we recently published [25]. See that protocol for additional details.

1. Reporter plasmids: Reporter plasmid with the HIV-1 frameshift stimulatory sequence (FSS) with or without an upstream 10× HA epitope tag can be acquired from Addgene (Plasmid #128610 or Plasmid #128611). From these plasmids, the multi-frame epitopes, FLAG (0 frame) and SunTag (−1 frame), and/or the human AlexX/XXLb1 (0/−1) open reading frames can be subcloned into another reporter plasmid with another FSS of interest (*see* **Note 1**).

2. Purified scFv-sfGFP: Standard protein expression and purification from *E. coli* should be performed. scFv-sfGFP can be purified using Ni-NTA-agarose (Qiagen) per the manufacturer's instructions. The eluted his-tagged scFv-sfGFP should be dialyzed in a HEPES-based buffer. The working stock should be ~1 mg/mL. The expression vector is available upon request. Alternatively, a mammalian expression plasmid system can be used to co-express scFv-sfGFP in cells via either transient transfection or direct bead loading of plasmid DNA (*see* **Note 2**).
3. Purified Cy3-conjugated anti-FLAG Fab: Prepare anti-FLAG Fab from anti-FLAG antibody with the Fab preparation kit. Label anti-FLAG Fab with Cy3 NHS ester, and then purify Cy3 anti-FLAG Fab with ultrafiltration. The working stock should be ~1 mg/mL.
4. Purified Halo-MCP: Standard protein expression and purification from *E. coli* should be performed. Halo-MCP can be purified using Ni-NTA-agarose (Qiagen) per the manufacturer's instructions. The working stock should be ~1 mg/mL. The expression vector is available upon request. Alternatively, a mammalian expression plasmid system can be used to co-express Halo-MCP in cells via either transient transfection or direct bead loading of plasmid DNA (*see* **Note 2**).

2.2 Tissue Culture

1. U-2 OS cells or other adherent cells such as HeLa or RPE-1 are recommended for efficient bead loading [26].
2. Phosphate-buffered saline (PBS).
3. Trypsin.
4. DMEM (+): Add 50 mL fetal bovine serum, 5 mL penicillin-streptomycin (10,000 U/mL), and 5 mL L-Glutamine (200 mM) to 500 mL Dulbecco's Modified Eagle Medium (DMEM, high glucose, no glutamine).

2.3 Imaging

1. Glass-bottom dishes (*see* **Note 3**).
2. 106 μm glass beads for bead loading purified proteins and/or plasmids (*see* **Note 2**): Sterilize glass beads in 2 M NaOH for 2 h. Wash the beads with Milli-Q water until the pH is neutral. Then rinse the beads with 100% ethanol. Keep the beads dry (*see* **Note 4**).
3. Opti-MEM.
4. Phenol-free DMEM (+).
5. Janelia Flour 646 HaloTag Ligands: Dissolve Janelia Flour 646 HaloTag Ligands (JF646-Halo) to 10 mM in DMSO for the stock solution. Dilute the stock solution to 0.2 mM in

DMSO for the working stock solution. Store at −20 °C (*see* **Note 5**).

6. Harringtonine: Dissolve Harringtonine to 3 mg/mL in DMSO as a 1000X working stock. Store at −20 °C for no greater than 1 month to prevent loss of potency. Mix well prior to use.
7. Puromycin: Dissolve puromycin dihydrochloride to 50 mg/mL in sterile H_2O.

3 Methods

3.1 Fab Generation and Dye Conjugation

1. Generate and purify Fab with the Pierce mouse IgG1 preparation kit per the manufacturer's instructions.
2. Incubate FLAG (anti DYKDDDDK mouse IgG_{2b} monoclonal) or HA (HA-7 IgG_1 mouse monoclonal) antibodies with ficin conjugated beads with 25 mM cysteine to generate their respective Fabs.
3. Isolate Fabs using NAb Protein A column.
4. To label Fabs with Cy3 or Alexa Fluor 488 (A488), use either the Cy3 N-hydroxysuccinimide ester or A488 tetrafluorophenyl ester by combining 100 μg of Fab (at ~1 mg/mL) combined with 10 μL of 1 M $NaHCO_3$, to a final volume of 100 μL (*see* **Note 6**).
5. Add either 2.66 μL of Cy3 or 5 μL of A488 directly to the 100 μL mixture. Incubate the reaction for 2 h at room temperature protected from light with end-over-end rotation.
6. Elute dye conjugated Fabs in PBS using PD-mini G-25 desalting column.
7. Concentrate dye conjugated Fabs to 1 mg/mL using Ultrafree 0.5 filter with a 10 k-cutoff.
8. Calculate the ratio of Fab/dye using the absorbance at 280 and 550 nm (for Cy3) or 495 nm (for A488); the extinction coefficient of Fab at 280 nm, $\mathcal{E}_{Fab}$; the extinction coefficient of dye, $\mathcal{E}_{dye}$; provided by the manufacturer; and the dye correction factor at 280 nm, CF, also provided by the manufacturer. Calculate the degree of labeling, DOL, using the following formula:

$$\mathrm{DOL} = \left(\frac{\mathcal{E}_{\mathrm{Fab}}}{\mathcal{E}_{\mathrm{dye}}}\right)\left(\frac{1}{\left(\frac{A_{280}}{A_{\mathrm{dye}}}\right) - \mathrm{CF}}\right) \tag{1}$$

Fab calculated with a DOL ~1 is recommended to be used in experiments.

3.2 MCP and scFv-sfGFP Purification

1. Express his-tagged MCP or scFv-sfGFP (see Materials) in a bacterial system.
2. Lyse bacteria in a PBS-based buffer with protease inhibitors.
3. Immobilize MCP or scFv-sfGFP on the Ni-NTA resin in the presence of 10 mM imidazole.
4. Wash the resin with 20 and 50 mM imidazole in PBS to remove non-specific immobilized proteins.
5. Elute MCP or scFv-sfGFP from the Ni-NTA resin using 300 mM imidazole in PBS.
6. Buffer exchange the MCP or scFv-sfGFP via dialysis into a HEPES-based buffer (10% glycerol, 25 mM HEPES pH 7.9, 12.5 mM $MgCl_2$, 100 mM KCl, 0.1 mM EDTA, 0.01% NP-40 detergent, and 1 mM DTT).
7. Snap-freeze ~1 mg/mL working aliquots in liquid nitrogen, and store at −80 °C.

3.3 Cell Culture, Transfection, and Bead-Loading

1. Culture U-2 OS cells in DMEM with 10% (v/v) FBS, 1 mM L-glutamine, and 1% (v/v) penicillin–streptomycin (*see* **Note** 7).
2. One day prior to imaging, passage U-2 OS cells at 70–90% confluency in a 35 mm MatTek chamber.
3. 1–2 h prior to imaging, transiently transfect the U-2 OS cells with reporter plasmid using the Lipofectamine LTX per the manufacturer's instructions, or bead-load the reporter plasmid with the purified probes (MCP and/or scFv) (*see* **Notes 2** and **8**).
4. Bead-loading involves the following generalized procedure and has been described comprehensively recently in video [26]:
   - (a) Combine 100 μg/mL of fluorescently labeled Fab, 250 μg/mL of purified scFv-sfGFP, and 33 μg/mL of purified MCP-HaloTag protein with PBS to a final volume of 4 μL.
   - (b) Aspirate and set aside the DMEM from the MatTek chamber.
   - (c) Pipette the 4 μL mixture on top of the cells.
   - (d) Sprinkle ~106 μm glass beads (Sigma Aldrich) evenly over the cells in a monolayer.
   - (e) Tap the chamber firmly yet carefully ~10 times on the cell culture hood bench top.
   - (f) Decant DMEM immediately back onto the cells such that the beads are gently pulled away from the cells.
   - (g) Allow the cells to recover by returning the chamber to the 37 °C incubator for at least 1 h.

(h) Optionally, DNA can also be bead-loaded with the other macromolecular probes (*see* **Note 2**). For each plasmid, approximately 1 μg of DNA is added to the initial 4 μL mix (the plasmid DNA working stock solution concentration should be close to 1 mg/mL).

5. Gently wash cells with phenol-red-free complete DMEM, and carefully wash away as many glass beads as possible.
6. Stain the MCP-HaloTag using 200 nM of JF646-HaloTag ligand, a cell permeable fluorogenic ligand [27].
7. Incubate for 30 min, and wash cells three times with 37 °C warmed phenol-red-free complete DMEM removing unconjugated fluorophores.
8. Immediately begin imaging experiments.

3.4 Single-Molecule Tracking Microscopy

1. Place live transfected/bead-loaded cells onto a stage top incubator set to a temperature of 37 °C and supplemented with 5% CO_2, and search for cells containing >10 RNAs in the cytoplasm (*see* **Note 9**).
2. To image and track single RNA with high fidelity, a HILO microscope is recommended [28], although a spinning disc confocal or light-sheet microscope with sensitive EMCCD or sCMOS cameras can also work.
3. To adequately track most RNA, image at a rate of approximately one cell volume every 1 or 2 s. For this, a TTL-triggered piezo stage is recommended to capture the entire cell volume as quickly as possible.
4. Image for at least 15 frames in all three channels to best detect, track, and distinguish translating from non-translating RNA.

3.5 Particle Tracking

1. Pre-process images using Fiji to make 2D maximum intensity projections from 3D images [29].
2. Detect the location of all RNA foci in each cell using a band-pass filter followed by an intensity threshold filter in the RNA channel. The threshold value should be set such that most pixels in RNA foci are above it, while most non-RNA pixels are below it. This will create a binarized image where RNA foci have a value of 1 and all other pixels have a value of 0. The positions of spots within this mask can be used to determine the centroids of each RNA in the cell.
3. Fit each detected RNA with a Gaussian to precisely determine its position and estimate the intensity in each channel. To facilitate this, extract an 11 × 11 grid of pixels (assuming each pixel is 130 nm) centered on each detected RNA (the size of

the pixel grid should fully cover the detected RNA) to a 2D Gaussian of the following form:

$$I(x,y) = I_{BG} + I e^{-\frac{(x-x_0)^2}{2\sigma_x^2} - \frac{(y-y_0)^2}{2\sigma_y^2}} \tag{2}$$

where I_{BG} is the background intensity, I is the detected RNA peak intensity in each channel, (σ_x, σ_y) are the fluorescence spread of the detected RNA in each channel, and (x_0, y_0) specify the location of the RNA.

4. Track RNA from frame to frame using a nearest-neighbor tracking algorithm such as TrackMate [30]. Hand-curate tracks to ensure they are accurate, eliminating tracks that are not clear due to cross-over of other tracks within dense RNA environments.

5. If necessary, perform a moving average of each movie to more accurately detect and identify frameshifting RNA (*see* **Note 10**).

3.6 Translation Intensity Measurements and Frameshift Identification

1. Measure the intensity of each translation signal (from the 0 and − 1 frame translation channels) using a disk-ring method, as shown in Fig. 2. Briefly, within an 11 × 11 pixel grid (assuming 130 nm/pixel) centered on each detected RNA, measure the average signal in a central circle of radius ~ 4 pixels (this central circle should be just large enough to cover each detected RNA). From this measured signal, subtract the background signal. The background signal can be estimated by measuring the average signal in a ring of pixels that surrounds the central circle. Although the 2D Gaussian fit described above can also be used to estimate signal intensities, we find the fits sometimes fail to accurately estimate near-zero

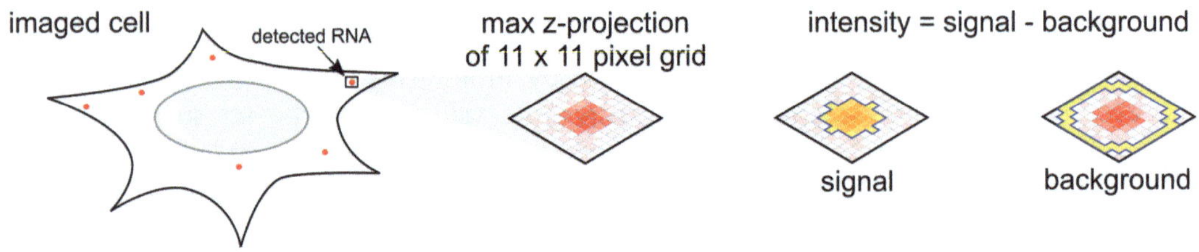

Fig. 2 Quantifying intensity signals at detected RNA. Schematic of the procedure to quantify the intensity of detected RNA spots. On the left, a cell is imaged, and an RNA is detected in the RNA channel (red dot with black square around it). From the detected spot coordinates, an 11 × 11 grid of pixels centered on the RNA spot is created. On the right, the intensity is calculated using a disk-ring method. The "signal" is the average pixel intensity in the central yellow disk of radius ~ 4 pixels (130 nm/pixel). The "background" is the average pixel intensity in the yellow ring surrounding the central disk. Intensity quantification is done for each channel collected. Note that if different channels are imaged using different cameras, the detection coordinates may need to be shifted from channel to channel to ensure the spot remains centered

intensities. The disk-ring mask method, on the other hand, ensures all measured signals are sensible, even when low.

2. Categorize each RNA foci as either non-frameshifting or frameshifting based on the signal intensities of the 0 (non-frameshifted) and − 1 (frameshifted) frames.

3.7 Confirm Translation Elongation in Both Open Reading Frames

1. To confirm active translation elongation, add puromycin at a final concentration of 100 μg/mL, and continuously image the entire cell volume every 30 s for an additional ~5 min. Translation signals should disappear quickly upon addition of puromycin (*see* **Note 11**).

3.8 Ribosome Run-off Experiments and Fits

1. Measure the average elongation rate by acquiring five pre-images of entire cell volumes every 60 or 120 s (*see* **Note 12**).
2. Treat cells with a final concentration of 3 μg/mL Harringtonine.
3. Immediately after addition of Harringtonine, continuously image the cell(s) for 30 more time points acquiring the full cell volume every 60 or 120 s. Depending on the length of the reporter open reading frame, this time can be extended or shortened.
4. Measure the intensity of all translation sites in each frame as described in Subheading 3.6 (Translation Intensity Measurements and Frameshift Identification).
5. Total the intensity of all translation sites in each frame (and in all cells) to produce the run-off curve of non-frameshifting sites and frameshifting sites. Normalize the curve so the total intensity starts off at ~1 (*see* **Note 13**).

   (i) Fit the curves to a linear regression to roughly estimate run-off times. The linear portion of the run-off decay begins when the normalized run-off intensity reaches a fraction f_0:

$$f_0 = L1/(L1 + L2/2) \tag{3}$$

where $L1$ is the length of untagged portion of the open reading frame and $L2$ is the length of the tagged portion (i.e., the length of the repeated epitopes), as described in [31].

   (ii) Estimate the run-off time R_T by interpolating the linear portion of the decay to background levels (*see* **Note 14**). The elongation rate k_{el} can be calculated as follows:

$$k_{el} = (L1 + L2/2)/R_T \tag{4}$$

4 Notes

1. The multi-frame epitopes, FLAG (0 frame) and SunTag (−1 frame), in frame with the human AlexX/XXLb1 [32, 33] (0/−1), are flanked by XbaI and NheI restriction sites. They can be subcloned in frame with another frameshift stimulatory sequence to monitor the 0 (FLAG) and − 1 (SunTag) frames or shifted by one nucleotide to monitor the +1 (FLAG) and 0 (SunTag) frames. Because frameshifting is a translation elongation phenomenon, there should be at least 23 codons upstream of the FSS to provide space for elongating ribosomes.
2. If the cell type used adheres well to coverslips, DNA plasmids and/or purified Fab, MCP, and scFv-sfGFP can all be bead-loaded simultaneously. This is a straightforward procedure that can be cheaper and often more efficient than transient transfection. Bead-loading also allows for more control over the concentration of loaded purified protein, which can be useful for optimizing experimental signal to noise. Details can be found in a separate video protocol we published [26]. For more sensitive cell types that do not adhere well to coverslips and tend to peel during bead loading, we recommend co-transfecting mammalian expression plasmids expressing the reporter and corresponding probes.
3. A 14 mm microwell MatTek glass-bottom dish is an ideal surface area for 4 μL of sample volume to load during bead-loading.
4. 106 μm glass beads are easily aerosolized and can be inhaled which may be harmful to the lungs. It is therefore recommended to only use the beads under a biosafety cabinet.
5. Minimize freeze-thaw cycles with JF646 HaloLigand by making working stock aliquots of 10 μL.
6. We recommend using fresh $NaHCO_3$ as decomposition can affect the labeling efficiency.
7. U-2 OS cells are used because of their flat morphology allowing for a minimal z-stack to capture the entire volume of the cell.
8. Lipofectamine reagent should be removed before bead-loading, and cells should be in fresh OPTI-MEM.
9. RNA should appear as distinct foci excited by ~637 nm laser if using JF646-HaloTag ligand (red spots in Fig. 1b). RNA can be co-localized with distinct foci excited by the ~561 nm laser (0 open reading frame green spots in Fig. 1b) or the ~488 nm laser (−1 open reading frame blue spots in Fig. 1b).
10. If more than one camera is being used to image in different colors, the various channels from each camera may not be perfectly aligned. To correct for this, images of multicolor

tetraspeck beads can be acquired prior to performing experiments [34]. From these images, the geometric transformation function that corrects for the shift in each channel can be determined and applied using Fiji plugins (such as Linear Stack Alignment with SIFT Multi-channel or TurboReg).

11. Signals typically disappear within a minute of puromycin addition if active elongation is occurring.
12. Laser powers should be lowered to minimize observable photobleaching. This will need to be determined empirically depending on the specific microscope setup.
13. The run-off curves should be normalized to the mean of the total intensity of the first 3–5 time points prior to treatment with Harringtonine.
14. Frameshifting translation sites may show a delayed response to Harringtonine which may be due to a queue of ribosomes upstream of the PRF. This will result in the same elongation rate but give insight into the length of the ribosomal queue upstream of the PRF.

Acknowledgments

We thank all members of the Stasevich and Munsky labs for helpful discussion and comments. This work was supported by the WM Keck Foundation and by a grant from the National Institutes of Health (R35GM119728 to TJS).

References

1. Morisaki T, Lyon K, DeLuca KF et al (2016) Real-time quantification of single RNA translation dynamics in living cells. Science 352: 1425–1429
2. Yan X, Hoek TA, Vale RD et al (2016) Dynamics of translation of single mRNA molecules in vivo. Cell 165:976–989
3. Wu B, Eliscovich C, Yoon YJ et al (2016) Translation dynamics of single mRNAs in live cells and neurons. Science 352:1430–1435
4. Wang C, Han B, Zhou R et al (2016) Real-time imaging of translation on single mRNA transcripts in live cells. Cell 165:990–1001
5. Pichon X, Bastide A, Safieddine A et al (2016) Visualization of single endogenous polysomes reveals the dynamics of translation in live human cells. J Cell Biol. jcb.201605024
6. Tanenbaum ME, Gilbert LA, Qi LS et al (2014) A protein-tagging system for signal amplification in gene expression and fluorescence imaging. Cell 159:635–646
7. Viswanathan S, Williams ME, Bloss EB et al (2015) High-performance probes for light and electron microscopy. Nat Methods 12: 568–576
8. Cialek CA, Koch AL, Galindo G et al (2020) Lighting up single-mRNA translation dynamics in living cells. Curr Opin Genet Dev 61:75–82
9. Zhao N, Kamijo K, Fox PD et al (2019) A genetically encoded probe for imaging nascent and mature HA-tagged proteins in vivo. Nat Commun 10:2947
10. Boersma S, Khuperkar D, Verhagen BMP et al (2019) Multi-color single-molecule imaging uncovers extensive heterogeneity in mRNA decoding. Cell 178:458–472.e19
11. Voigt F, Zhang H, Cui XA et al (2017) Single-molecule quantification of translation-dependent association of mRNAs with the endoplasmic reticulum. Cell Rep 21:3740–3753

12. Goldman DH, Livingston NM, Movsik J et al (2021) Live-cell imaging reveals kinetic determinants of quality control triggered by ribosome stalling. Mol Cell 81:1830–1840.e8
13. Cialek CA, Galindo G, Morisaki T et al (2022) Imaging translational control by Argonaute with single-molecule resolution in live cells. Nat Commun 13:3345
14. Moon SL, Morisaki T, Khong A et al (2019) Multicolour single-molecule tracking of mRNA interactions with RNP granules. Nat Cell Biol 21:162–168
15. Mateju D, Eichenberger B, Voigt F et al (2020) Single-molecule imaging reveals translation of mRNAs localized to stress granules. Cell 183: 1801–1812.e13
16. Koch A, Aguilera L, Morisaki T et al (2020) Quantifying the dynamics of IRES and cap translation with single-molecule resolution in live cells. Nat Struct Mol Biol 27:1095–1104
17. Lyon K, Aguilera LU, Morisaki T et al (2019) Live-cell single RNA imaging reveals bursts of translational frameshifting. Mol Cell 75:172–183.e9
18. Atkins JF, Loughran G, Bhatt PR et al (2016) Ribosomal frameshifting and transcriptional slippage: from genetic steganography and cryptography to adventitious use. Nucleic Acids Res gkw530
19. Wang X, Xuan Y, Han Y et al (2019) Regulation of HIV-1 gag-pol expression by shiftless, an inhibitor of programmed -1 ribosomal frameshifting. Cell 176:625–635.e14
20. Firth AE, Jagger BW, Wise HM et al (2012) Ribosomal frameshifting used in influenza A virus expression occurs within the sequence UCC_UUU_CGU and is in the +1 direction. Open Biol 2:120109
21. Dinman JD (2009) Programmed −1 ribosomal frameshifting in SARS coronavirus. In: Molecular biology of the SARS-coronavirus, pp 63–72
22. Bhatt PR, Scaiola A, Loughran G et al (2021) Structural basis of ribosomal frameshifting during translation of the SARS-CoV-2 RNA genome. Science 372:1306–1313
23. Atkins JF, O'Connor KM, Bhatt PR et al (2021) From recoding to peptides for MHC class I immune display: enriching viral expression, virus vulnerability and virus evasion. Viruses 13:1251
24. Grentzmann G, Ingram JA, Kelly PJ et al (1998) A dual-luciferase reporter system for studying recoding signals. RNA N Y N 4: 479–486
25. Koch AL, Morisaki T, Stasevich TJ (2021) A multi-color bicistronic biosensor to compare the translation dynamics of different open reading frames at single-molecule resolution in live cells. Bio-Protoc 11:e4096
26. Cialek CA, Galindo G, Koch AL et al (2021) Bead loading proteins and nucleic acids into adherent human cells. J Vis Exp 62559
27. Grimm JB, English BP, Chen J et al (2015) A general method to improve fluorophores for live-cell and single-molecule microscopy. Nat Methods 12:244
28. Tokunaga M, Imamoto N, Sakata-Sogawa K (2008) Highly inclined thin illumination enables clear single-molecule imaging in cells. Nat Methods 5:159–161
29. Schindelin J, Arganda-Carreras I, Frise E et al (2012) Fiji: an open-source platform for biological-image analysis. Nat Methods 9: 676–682
30. Tinevez J-Y, Perry N, Schindelin J et al (2017) TrackMate: an open and extensible platform for single-particle tracking. Methods 115:80–90
31. Morisaki T, Stasevich TJ (2018) Quantifying single mRNA translation kinetics in living cells. Cold Spring Harb Perspect Biol 10:a032078
32. Abramowitz J, Grenet D, Birnbaumer M et al (2004) XLαs, the extra-long form of the α-subunit of the Gs G protein, is significantly longer than suspected, and so is its companion Alex. Proc Natl Acad Sci 101:8366–8371
33. Aydin C, Aytan N, Mahon MJ et al (2009) Extralarge XLαs (XXLαs), a variant of stimulatory G protein α-subunit (Gsα), is a distinct, membrane-anchored GNAS product that can mimic Gsα. Endocrinology 150:3567–3575
34. Lowe DG (2004) Distinctive image features from scale-invariant keypoints. Int J Comput Vis 60:91–110

Chapter 10

Benzo[*c,d*]Indole-Quinoline-Based Deep-Red Emissive Probes for Live-Cell Imaging of Nucleolar RNA

Michiyuki Suzuki, Yusuke Sato, and Seiichi Nishizawa

Abstract

Small molecular weight fluorescent probes capable of binding to RNAs have been powerful tools for understanding the intracellular behaviors of RNAs. In this chapter, we describe the fluorescence imaging of nucleolar RNA in living cells using deep-red emissive probes with benzo[*c,d*]indole-quinoline (BIQ) monomethine cyanine scaffolds. These probes feature a significant fluorescence "off-on" ability upon binding to RNAs (>100-fold) in the deep-red spectral region ($\lambda_{em} > 650$ nm). In addition, they have many advantages for fluorescence sensing of RNAs and nucleolar RNA imaging, such as longer emission wavelength, higher photostability, and better counterstaining compatibility for live cell imaging, compared to a commercially available RNA-binding probe, SYTO RNA select.

Key words Nucleolar RNA, Live-cell imaging, Deep-red fluorescence, Cyanine dyes

1 Introduction

With increasing knowledge about the diverse role of RNAs within cells, analytical methods to visualize intracellular RNAs have gained much attention for the study of RNA functions in the biological process [1]. Most of the RNA in the nucleus is localized to the nucleolus that plays a crucial role in ribosomal RNA transcription, processing, and ribosome assembly. The nucleolus has been considered as an important target for diagnostic and therapeutic applications [2, 3]. In this context, small molecular weight fluorescent probes targeting nucleolar RNA have emerged as useful tools due to their favorable abilities for cell imaging experiments, such as versatility and easy-to-use properties. SYTO RNA select is the only commercially available probe that can be applied to the imaging of nucleolar RNA in cells. When free in solution, SYTO RNA select shows negligible emission, but its fluorescence is greatly enhanced upon binding to RNAs. However, the usefulness of SYTO RNA for the cell imaging has been severely limited

Shin-ichi Sato (ed.), *Live-Cell RNA Imaging: Methods and Protocols*, Methods in Molecular Biology, vol. 2875,
https://doi.org/10.1007/978-1-0716-4248-1_10,

[4]. This probe has problems of low photostability and the relatively short emission wavelength (λ_{em} = 530 nm) that suffers interference from the cellular autofluorescence. More seriously, SYTO RNA select is hard to be applied for the nucleolar RNA imaging in living cells [4–6], due to poor cell membrane permeability. The chemical structure of SYTO RNA select has not been described yet, which renders it difficult to improve the properties based on SYTO RNA select. Hence, much effort has been devoted to the development of new chemical scaffolds to overcome the above drawbacks of SYTO RNA select.

Recently, we developed a new nucleolar RNA-targeting fluorescent probe with the improved photophysical properties and compatibility toward live-cell imaging based on benzo[*c,d*]indole–quinoline (BIQ) monomethine cyanine scaffold [6]. BIQ represents one of rare examples of nucleolar RNA-imaging probes having deep-red emission range (λ_{em} > 650 nm). BIQ is almost non-fluorescent due to nonradioactive energy loss by the free rotation around the monomethine bridge between benzo[*c,d*]indole and quinoline rings, similar to various monomethine cyanine dyes. The binding to target RNA causes the significant increase in the emission intensity of BIQ (more than 100-fold) in the deep-red spectral region (λ_{em} = 657 nm), which can be explained by the restricted rotation of BIQ by intercalation into the base pairs in the RNA. The light-up signaling ability of BIQ for RNA is outstanding compared to coumarin-pyronin (CP, λ_{em} = 658 nm) [7] and Nile blue derivative (NBE, λ_{em} = 705 nm) [8] that were previously developed as deep-red emissive small probes for nucleolar RNA imaging in the living cells (less than fivefold). This arises from the torsional flexibility of the two heterocycles around the central bridge in the BIQ structure, the key for the light-up property, in sharp contrast to CP and NBE with rigid chemical structures. Another point of significance in the fluorescence property of BIQ is much higher photostability over SYTO RNA select. BIQ has good cell plasma and nuclear membrane permeability in living cells, and the nucleolus can be stained upon binding to nucleolar RNA by 20 min incubation of 5.0 μM BIQ. We demonstrated negligible cytotoxicity at the concentration used in typical live-cell imaging experiments. BIQ is thus the most useful candidate as a deep-red probe suitable for the imaging of nucleolar RNA in living cells with a view toward the analysis of nucleolar dynamics, such as nucleolar RNA processing during mitosis and upon treatment with anticancer drugs. We also developed a BIQ derivative carrying an amino group-terminated side chain, BIQ-NH_2, with the improved binding affinity for RNA due to the favorable electrostatic interactions compared to the parent BIQ [9]. BIQ-NH_2 exhibited the enhanced detection sensitivity of nucleolar RNA in living cells, where the nucleolar RNA imaging can be achieved even at the concentration as low as 500 nM.

Here we describe our procedures for the synthesis and fluorescence imaging of nucleolar RNA in living cells using BIQ and BIQ-NH_2. We also describe a protocol for co-staining experiments with other organelle-selective probes. In addition, a protocol for deoxyribonuclease (DNase) and ribonuclease (RNase) digestion experiments was shown for the assessment of their RNA-binding properties inside the cells. Taken together, we hope that these methods will be useful for the fluorescence imaging of nucleolar RNA with a view toward the analysis of nucleolar dynamics in living cell.

2 Materials

2.1 Chemicals

1. 1,4-Dioxane.
2. 3-(Boc-amino)propyl bromide.
3. CH_3CN.
4. $CHCl_3$.
5. DMF.
6. DMSO.
7. HCl (4 M).
8. K_2CO_3.
9. Lawesson's reagent.
10. MeOH.
11. Na_2SO_4.
12. Triethylamine (TEA).
13. Benzo[*c,d*]indol-2(1H)-one.
14. Brine.
15. Diethyl ether.
16. Ethyl acetate.
17. Iodomethane.
18. Lepidine.

2.2 Cell Experiment

1. MCF7.
2. Culture medium.

    RPMI1640 (+ phenol red, + L-glutamine) supplemented with 10% (v/v) fetal bovine serum.

    (FBS) and 2% (v/v) penicillin–streptomycin.
3. D-PBS (−).
4. Trypsin/EDTA.
5. HBSS (+).

6. Methanol.
7. DNase I (RNase-free) and 10 × DNase reaction buffer.
8. RNase A (DNase-free).
9. BIQ solution (400 μL of culture medium containing 5.0 μM BIQ).
10. BIQ-NH_2 solution (400 μL of culture medium containing 500 nM BIQ-NH_2).
11. Nuclease solution; DNase reaction buffer containing DNase (100 U/mL) and RNase A solution diluted with D-PBS (20 μg/mL).

2.3 Consumable

1. 100 mm cell culture dish.
2. Cell-counting slide.
3. Eight-chambered glass plate.

2.4 Equipment

1. NMR Bruker Advance III 500.
2. ESI-MS JEOL JMS-T100CS.
3. Rotary evaporator.
4. Milli-Q Synthesis A10 system.
5. BioPak filter.
6. Automated cell counter.
7. Delta Vision Elite Imaging System.

3 Methods

3.1 Synthesis of BIQ and BIQ-NH_2

The synthetic scheme of BIQ and BIQ-NH_2 is shown in Fig. 1.

3.2 Synthesis of 1-Methylbenzo[c,d] Indol-2(1H)-One (2)

1. Dissolve benzo[*c*,*d*]indol-2(1H)-one (**1**) (1.68 g, 9.93 mmol) and K_2CO_3 (2.74 g, 19.8 mmol) in DMF (30 mL).
2. Add 0.73 mL of iodomethane (11.8 mmol), and stir at 38 °C under nitrogen atmosphere for 2 days.
3. After cooling to room temperature, add 50 mL of deionized water.
4. Extract the mixture with ethyl acetate (75 mL × 2).
5. Wash organic layer with brine and dry over Na_2SO_4.
6. Remove the solid by filtration, and concentrate the filtrate under reduced pressure by rotary evaporator.
7. Purify the crude product by flash silica gel chromatography employing $CHCl_3$/MeOH gradient system to yield the purified **2** as a yellow solid (1.07 g, 59%) (*see* **Note 1**).

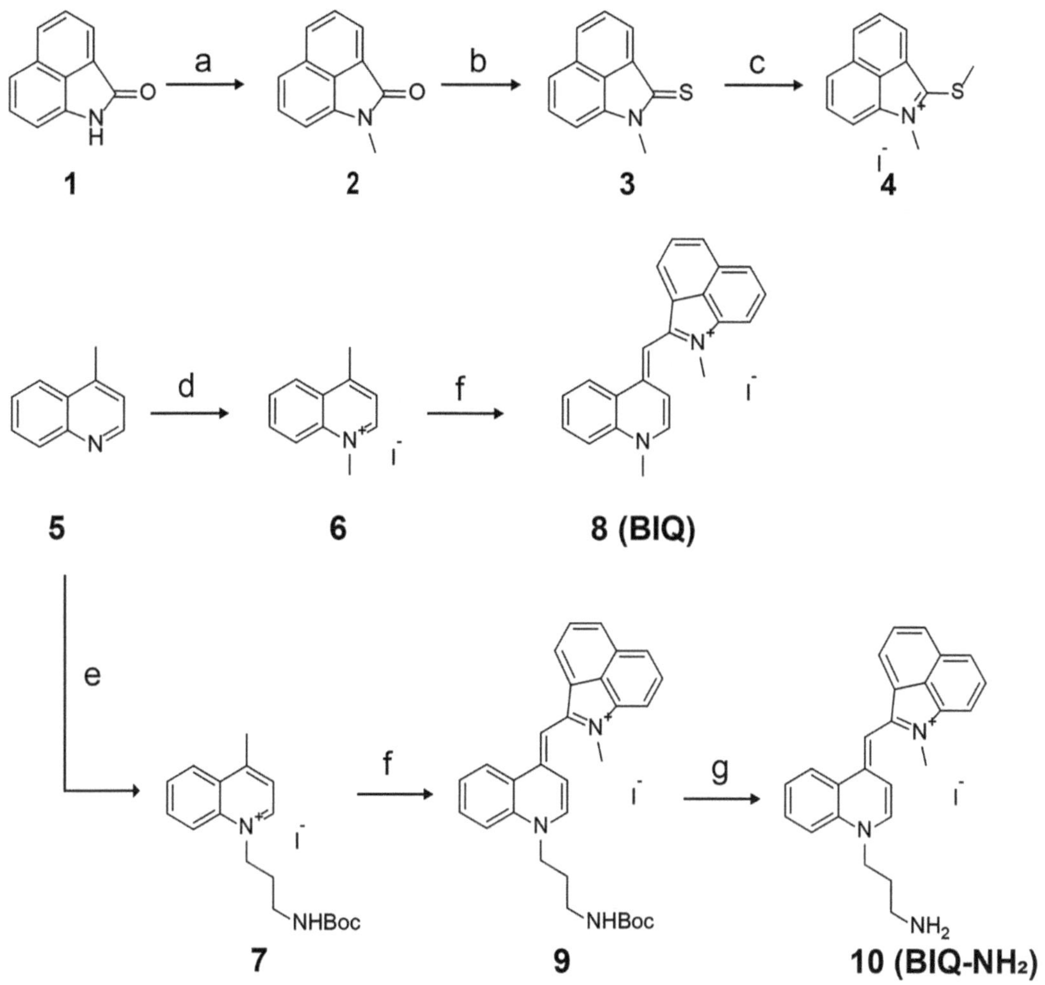

Reagents and conditons

a) CH_3I, K_2CO_3, DMF, 38 °C, 2 days
b) Lawesson' s reagent, 1,4 dioxane, reflux, overnight
c) CH_3I, reflux, 8 h
d) CH_3I, CH_3CN, reflux, 8 h
e) Br NHBoc, CH_3CN, reflux, 24 h
f) Compound **4**, TEA, CH_3CN, reflux, 1 h
g) HCl, 1,4 dioxane, r.t., 1h

Fig. 1 Synthetic scheme of BIQ (compound **8**) and BIQ-NH_2 (compound **10**)

3.3 Synthesis of 1-Methylbenzo[c,d] Indol-2(1H)-Thione (3)

1. Dissolve 1-methylbenzo[*c,d*]indol-2(1H)-one (**2**) (1.07 g, 5.8 mmol) and Lawesson's reagent (4.72 g, 11.6 mmol) in 1,4-dioxane (40 mL), and stir at 100 °C under nitrogen atmosphere for 18 h.

2. After cooling to room temperature, collect the solid by filtration, and wash with 1,4-dioxane.
3. Purify the crude product by silica gel chromatography employing $CHCl_3$ to yield the purified **3** as an orange solid (0.77 g, 66.6%) (*see* **Note 2**).

3.4 Synthesis of 1-Methyl-2-(Methylthio) Benzo[c,d]Indole-1-Ium (4)

1. Stir the mixture of 1-methylbenzo[*c,d*]indol-2(1H)-thione (**3**) (0.77 g, 3.7 mmol) and iodomethane (3.37 mL, 52 mmol) at 40 °C under nitrogen atmosphere for 8 h.
2. After cooling to room temperature, remove the excess iodomethane under reduced pressure by rotary evaporator.
3. Add 50 mL of diethyl ether and sonicate the mixture for 30 min.
4. Collect the solid by filtration to yield the crude **4** as an orange solid, verified by ESI-MS. This was used for next step without further purification.

3.5 Synthesis of 1,4-Dimethylquinolin-1-Ium Iodide (6)

1. Stir the mixture of lepidine (**5**) (1.4 mL, 10.6 mmol) and iodomethane (0.1 mL, 11.2 mmol) in 5 mL of CH_3CN at room temperature under nitrogen atmosphere for 18 h.
2. Add excess diethyl ether and collect the solid by filtration. Wash with diethyl ether to yield purified **6** as a white solid (2.55 g, 85%) (*see* **Note 3**).

3.6 Synthesis of 1-(3-((Ethoxycarbonyl) Amino)Propyl)-4-Methylquinolin-1-Ium (7)

1. Stir the mixture of lepidine (**5**) (0.5 mL, 37.9 mmol) and 3-(boc-amino)propyl bromide (223 mg, 0.94 mmol) at 100 °C under nitrogen atmosphere for 6 h.
2. After cooling to room temperature, add 10 mL of deionized water.
3. Extract the mixture with diethyl ether (10 mL × 2).
4. Evaporate the water layer to yield the crude 7 as a purple solid, verified by ESI-MS.

This was used for next step without further purification.

3.7 Synthesis of BIQ (8)

1. Dissolve 1-methyl-2-(methylthio) benzo[*c,d*]indole-1-ium (**4**) (194 mg, 0.57 mmol) and 1,4-dimethylquinolin-1-ium iodide (**6**) (258 mg, 0.91 mmol) in 12 mL of CH_3CN.
2. Add 0.5 mL of TEA and heat the mixture to reflux under nitrogen atmosphere for 1 h.
3. After cooling to room temperature, add excess diethyl ether.
4. Collect the solid by filtration and wash with deionized water.
5. After dissolving the obtained solid in MeOH, add the excess diethyl ether.

6. Collect the solid by filtration to yield the purified BIQ as a blue solid (82.2 mg, 32%) (*see* **Note 4**).
7. Prepare the stock solution of BIQ by dissolving the obtained solid in MeOH, and keep it at refrigerator (5 °C) in the dark.

3.8 Synthesis of (E)-2-((1-(3-((Ethoxycarbonyl)Amino)Propyl)Quinolin-4(1H)-Ylidene)Methyl)-1-Methyl Benzo[c,d]Indol-1-Ium (9)

1. Dissolve 1-methyl-2-(methylthio)benzo[*c,d*]indole-1-ium (**4**) (45 mg, 0.13 mmol) and 1-(3-((ethoxycarbonyl)amino)propyl)-4-methylquinolin-1-ium (7) (50 mg, 0.13 mmol) in 3 mL of CH_3CN.
2. Add 0.13 mL of TEA, and heat the mixture to reflux under nitrogen atmosphere for 1 h.
3. After cooling to room temperature, concentrate the mixture in reduced pressure by rotary evaporator.
4. Purify the crude compound by silica gel chromatography employing $CHCl_3$/MeOH gradient to yield purified **9** as a blue solid (30 mg, 42%) (*see* **Note 5**).

3.9 Synthesis of BIQ-NH_2 (10)

1. Dissolve (E)-2-((1-(3-((ethoxycarbonyl)amino)propyl)quinolin-4(1H)-ylidene)methyl)-1-methyl benzo[cd]indol-1-ium (**9**) (30 mg, 0.0546 mmol) in 3 mL of 1,4-dioxane.
2. Add 1.5 mL of 4 M HCl and stir the mixture at room temperature for 1 h.
3. Concentrate the mixture under reduced pressure by rotary evaporator.
4. After dissolving the obtained residue in 4 mL of $CHCl_3$, add excess diethyl ether.
5. Collect the solid by filtration, and wash with diethyl ether to yield purified BIQ-NH_2 as a blue solid (0.67 mg, 3.6%) (*see* **Note 6**).
6. Prepare the stock solution of BIQ-NH_2 by dissolving the obtained solid in DMSO, and keep it at refrigerator (5 °C) in the dark.

3.10 Imaging of Living MCF7 Cells by BIQ with Hoechst 33342 Co-Staining

1. Warm all mediums and buffers to 37 °C before use.
2. Prepare 70 ~ 80% confluent MCF7 cells in a 100 mm cell culture dish (*see* **Note** 7).
3. Aspirate the medium from the dish and wash with 5 mL of D-PBS (−) twice.
4. Add 2 mL of trypsin/EDTA solution, and incubate the dish at 37 °C in a 5% CO_2 incubator for 2 min.
5. Gently tap the edge of the dish to detach the cells from the dish.
6. Add 2 mL of medium and mix by pipetting slowly.

7. Add the 20 μL of the obtained cell suspension to cell-counting slide, and count the number of cells using the automated cell counter.
8. Add 400 μL of medium to eight-chambered glass plate.
9. Add the cell suspension (5.0×10^3 cells) to the plate containing the medium, and mix by pipetting slowly (*see* **Note 8**).
10. Incubate the plate at 37 °C in a 5% CO_2 incubator for 24 h.
11. After aspirating the medium, add 400 μL of BIQ solution.
12. Incubate at 37 °C in a 5% CO_2 incubator for 20 min.
13. Aspirate the BIQ solution.
14. For the co-staining, repeat **steps 11–13** using the medium containing Hoechts 33,342 solution.
15. After washing with 400 μL of HBSS (+) twice (*see* **Note 9**), add 400 μL of HBSS (+) for the microscopic analysis.
16. Turn on the delta Vision Elite Imaging System at least 30 min before starting imaging.
17. Choose the proper object lens, and add a drop of an appropriate immersion oil to the object lens (*see* **Note 10**).
18. Put the eight-chambered glass plate on the pre-warmed stage.
19. Confirm that the object lens is connected with the eyepiece.
20. Focus the cells by eyepiece using DIC filter.
21. Change the connection of the object lens from the eyepiece to the screen.
22. Check the image, and decide the acquiring condition including the excitation and emission filters suitable for the probes (*see* **Note 11**).
23. Acquire the image.
24. Figure 2 shows fluorescence images of living MCF7 co-stained by BIQ and Hoechst 33342, a blue emissive DNA staining dye. From the comparison with the nucleus staining by Hoechst 33342, it was found that strong emission of BIQ was observed in nucleolus. This resulted from the binding of BIQ to nucleolar RNA as confirmed by the nuclease digestion experiments (Subheading 3.4). Meanwhile, the emission observed in the cytoplasm can be explained by the accumulation of BIQ in the mitochondria because it was well overlapped with the staining pattern with mitochondria-selective probe [6]. These results indicate that BIQ has a superior plasma and nuclear membrane permeability and can stain nucleolar RNA in living cells. Furthermore, BIQ has good counterstaining compatibility with Hoechst 33342 (*see* **Note 12**).

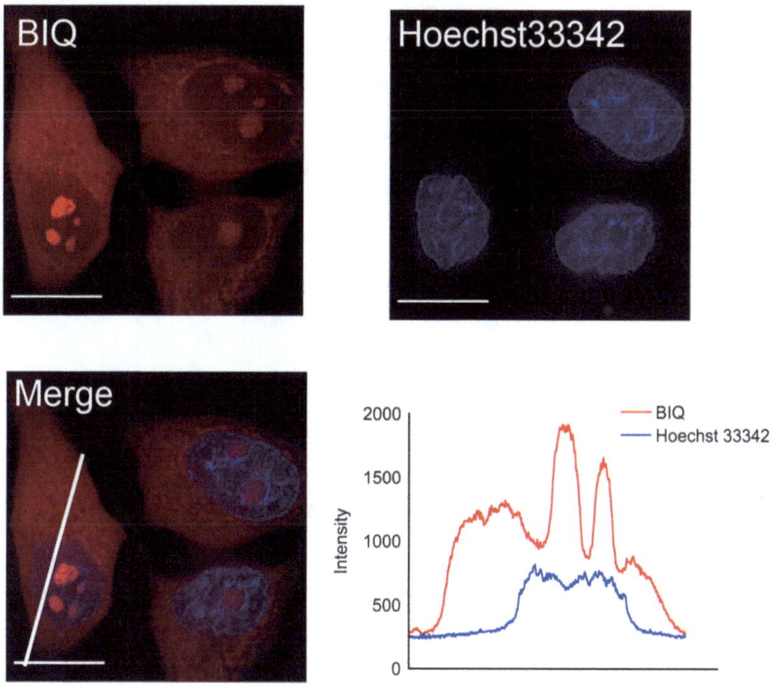

Fig. 2 Fluorescence images of living MCF7 stained by BIQ (5.0 μM) and Hoechst 33342 (176 nM). Fluorescence intensity profiles along the white line are also shown. Scale bar: 15 μm. The emission of BIQ and Hoechst 33342 was observed using the Cy5 filter set (excitation 632/22; emission 679/34) and DAPI filter set (excitation 390/18; emission 435/48), respectively. %T, 100 (BIQ) and 32 (Hoechst 33342). Exposure time, 1.0 s (BIQ) and 0.1 sec (Hoechst 33342)

3.11 Use of BIQ-NH_2 for the Improved Live-Cell Imaging of Nucleolar RNA

The procedure is the same as that shown in Subheading 3.2. BIQ-NH_2 was designed by introducing the primary amino group into the quinolone ring of BIQ through a propyl spacer [9]. BIQ-NH_2 features larger light-up response for RNA compared to BIQ due to the enhanced binding affinity while keeping the deep-red emissive property as well as higher photostability compared to SYTO RNA select. The fluorescence quantum yield of BIQ-NH_2 in the bound state with RNA is 2.3-fold higher than that of BIQ (ϕ_{bound}, BIQ–NH_2, 0.020; BIQ, 0.0085). This results in the stronger emission for nucleolar RNA by more than twofold over BIQ under the identical measurement condition [9]. The staining of nucleolar RNA in living cells is achieved even at the concentration of BIQ-NH_2 as low as 500 nM (Fig. 3).

3.12 Nuclease Digestion Experiments in Fixed-Permeabilized Cells

1. After **steps 1–10** shown in Subheading 3.2, aspirate the medium from the plate, and add 400 μL of chilled methanol.
2. Incubate at room temperature for 1 min and aspirate the methanol.
3. Add 100 μL of Triton X-100 solution and incubate at room temperature for 2 min.

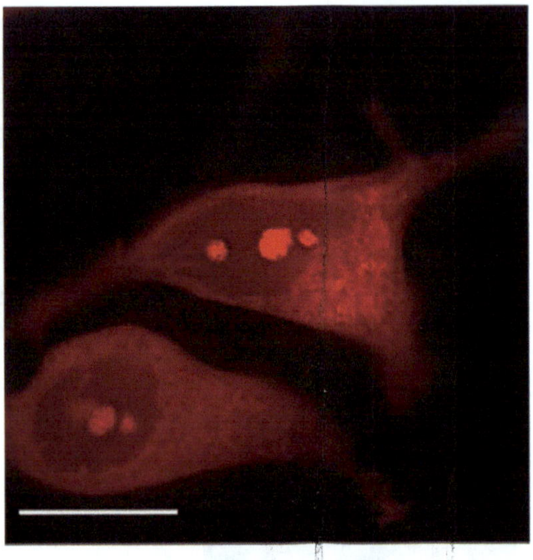

Fig. 3 Fluorescence image of living MCF7 stained by BIQ-NH_2 (500 nM). Scale bar: 15 μm. Cy5 filter set (excitation 632/22; emission 679/34) was used for the observation of BIQ-NH_2 emission. %T, 50; exposure, 0.5 s

4. Aspirate the Triton X-100 solution and wash with 400 μL of D-PBS (−) twice (*see* **Note 13**).
5. Add 400 μL of BIQ solution, and incubate at 37 °C in a 5% CO_2 incubator for 20 min.
6. Aspirate the BIQ solution and add 100 μL of nuclease solution or 100 μL of D-PBS (for control) (*see* **Notes 14** and **15**).
7. Incubate at 37 °C in a 5% CO_2 incubator for 3 h.
8. Aspirate the nuclease solution and wash with D-PBS (−) twice.
9. After adding 400 μL of D-PBS (−), image the cells using the fluorescence microscopy, as shown in **steps 16–23** in Subheading 3.2.
10. As an example of the results obtained by the nuclease digestion experiments, we showed the fluorescence image of fixed-permeabilized MCF7 cells stained by BIQ before (control) and after treatment of RNase or DNase (Fig. 4). The emission of BIQ is observed in the nucleolus for the control condition as observed for living cells (Fig. 2). The emission in the nucleolus remains almost unaffected upon DNase treatment. In contrast, RNase digestion results in the disappearance of BIQ emission in the nucleolus. These results clearly showed that strong emission of BIQ at the nucleolus originated from the binding to nucleolar RNA, such as ribosomal RNAs.

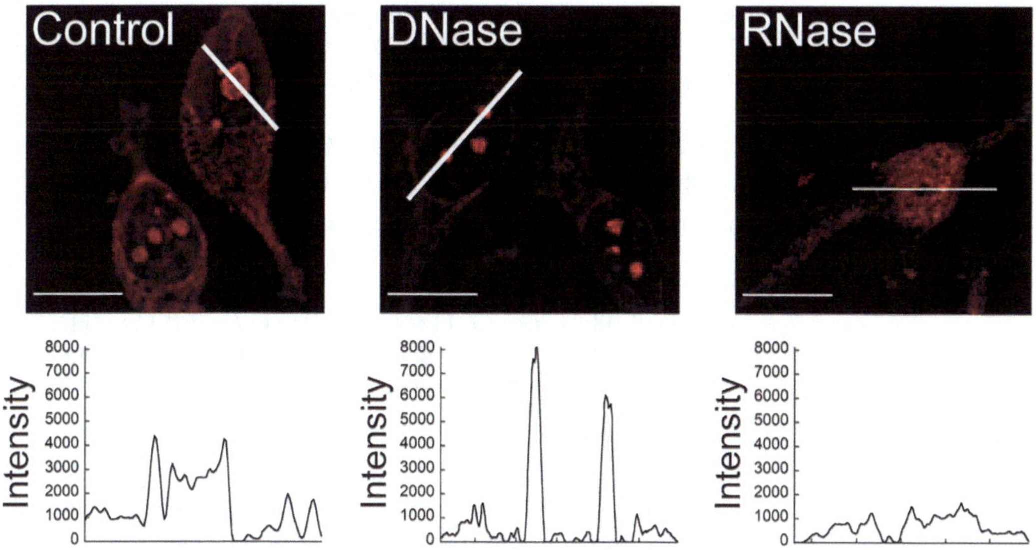

Fig. 4 Fluorescence image of fixed-permeabilized MCF7 cells stained with BIQ (5.0 μM) before (control) and after treatment with RNase or DNase. Fluorescence intensity profiles along the white line are also shown. Scale bar, 15 μm. Cy5 filter set (excitation 632/22; emission 679/34) was used for the observation of BIQ emission. %T, 100; exposure, 1.0 s

4 Notes

1. Characterization data of compound **2**.

   1H-NMR (500 MHz, DMSO-d_6) δ 8.17 (d, *J* = 8.2 Hz, 1H), 8.01 (d, *J* = 7.0 Hz, 1H), 7.80 (t, *J* = 7.5 Hz, 1H), 7.59 (d, *J* = 8.2 Hz, 1H), 7.52–7.47 (m, 1H), 6.98 (d, *J* = 7.0 Hz, 1H).

2. Characterization data of compound **3**.

   1H-NMR (500 MHz, $CDCl_3$) δ 8.23 (d, *J* = 7.3 Hz, 1H), 8.05 (d, *J* = 7.9 Hz, 2H), 7.74–7.63 (m, 3H), 7.50 (t, *J* = 7.6 Hz, 1H), 7.09 (d, *J* = 7.0 Hz, 1H), 1.56 (s, 4H).

3. Characterization data of compound **6**.

   1H-NMR (500 MHz, DMSO-d_6) δ 9.33–9.36 (m, 1H), 8.55 (dd, *J* = 8.4, 1.1 Hz, 1H), 8.49 (d, *J* = 8.8 Hz, 1H), 8.26–8.32 (m, 1H), 8.04–8.09 (m, 2H), 4.58 (s, 3H), 3.01 (s, 3H).

4. Characterization data of compound **8** (BIQ).

   H-NMR (700 MHz, DMSO-d_6): δ 8.87 (d, *J* = 4.7 Hz, 1H), 8.75 (d, *J* = 7.3 Hz, 1H), 8.62 (m, 1H), 8.48 (m, 1H), 8.30 (d, *J* = 9.0 Hz, 1H), 8.17 (t, *J* = 7.3 Hz, 1H), 8.04 (d, *J* = 7.7 Hz, 1H), 7.92 (t, *J* = 7.5 Hz, 1H), 7.54 (m, 3H), 7.21 (m, 1H), 6.79 (m, 1H), 4.40 (s, 3H), 3.73 (s, 3H).

5. Characterization data of compound **9**.

   1H-NMR (500 MHz, DMSO-d_6) δ 8.77 (d, J = 8.5 Hz, 1H), 8.35 (d, J = 8.8 Hz, 1H), 8.14 (t, J = 7.8 Hz, 1H), 7.90 (t, J = 7.8 Hz, 1H), 7.55 (d, J = 5.8 Hz, 2H), 7.25 (d, J = 7.6 Hz, 1H), 7.07–7.09 (M, 1H), 4.81 (T, J = 7.5 Hz, 2H), 3.50 (S, 1H), 3.11 (d, J = 6.7 Hz, 2H), 2.36 (S, 1H)
6. Characterization data of compound **10** (BIQ-NH_2).

   1H-NMR (500 MHz, DMSO-d_6) δ 8.91 (s, 1H), 8.78 (d, J = 8.2 Hz, 1H), 8.45 (d, J = 8.5 Hz, 1H), 8.16 (t, J = 7.8 Hz, 4H), 8.08 (d, J = 7.6 Hz, 1H), 7.91 (t, J = 7.8 Hz, 1H), 7.56 (s, 2H), 7.26 (d, J = 6.4 Hz, 1H), 4.93 (t, J = 7.6 Hz, 2H), 3.50 (s, 2H), 2.24–2.30 (m, 2H), 1.23 (s, 3H).
7. Do not use more than 90% confluent cells.
8. To acquire good images, cells for imaging should be 30–40% confluent in each well. Please adjust the number of cells to be added according to the growth rate of the cells.
9. Gentle mixing by pipetting is of importance in order to avoid the cell detaching.
10. If the lens in your system does not require the oil, you can skip this step.
11. The condition for acquiring the image should be carefully determined in order to avoid the saturation of the probe emission for the quantitative analysis.
12. BIQ was also shown to have good counterstaining compatibility with MitoTracker orange and Cell Navigator Lysosome staining kit [6].
13. It is important to wash completely the Triton-X solution for acquiring the good images.
14. In order not to lose the nuclease activity, it is recommended to avoid the freeze-thaw of nuclease solution more than three times.
15. If cells are not completely immersed in solution, add more 100 μL of the nuclease solution.

References

1. Suseela YV, Narayanaswamy N, Pratihar S, Govindaraju T (2018) Far-red fluorescent probes for canonical and non-canonical nucleic acid structures: current progress and future implications. Chem Soc Rev 47:1098–1131
2. Lindström MS, Jurada D, Bursac S, Orsolic I, Bartek J, Volarevic S (2018) Nucleolus as an emerging hub in maintenance of genome stability and cancer pathogenesis. Oncogene 37: 2351–2366
3. Weeks SE, Metje BJ, Samant RS (2019) The nucleolus: a central response hub for the stressors that drive cancer progression. Cell Mol Life Sci 76:4511–4524
4. Li Q, Kim Y, Namm J, Kulkani A, Rosania GR, Ahn Y-H, Chang Y-T (2006) RNA-selective, live cell imaging probes for studying nuclear structure and function. Chem Biol 13:615–623
5. Cao C, Wei P, Li R, Zhong Y, Li X, Xue F, Shi Y, Yi T (2019) Ribosomal RNA-selective light-up

fluorescent probe for rapidly imaging the nucleolus in live cells. ACS Sens 4:1409–1416

6. Yoshino Y, Sato Y, Nishizawa S (2019) Deep-red light-up signaling of benzo[c,d]indole–quinoline monomethine cyanine for imaging of nucleolar RNA in living cells and for sequence-selective RNA analysis. Anal Chem 91:14254–14260
7. Zhou B, Liu W, Zhang H, Wu J, Liu S, Xu H, Wang P (2015) Imaging nucleolar RNA in living cells using a highly photostable deep-red fluorescent probe. Biosens Bioelectron 68:189–196
8. Yao Q, Li H, Xian L, Xu F, Xia J, Fan J, Du J, Wang J, Peng X (2018) Differentiating RNA from DNA by a molecular fluorescent probe based on the "door-bolt" mechanism biomaterials. Biomaterials 177:78–87
9. Sato Y, Igarashi Y, Suzuki M, Higuchi K, Nishizawa S (2021) Deep-red fluorogenic cyanine dyes carrying an amino-group-terminated side chain for improved RNA detection and nucleolar RNA imaging. RSC Adv 11:35436–35439

Chapter 11

Direct Visualization of Hepatitis C Virus RNA in Living Cells and Mice

Shi Kuang, Jiao-Na Han, Jiaheng Zhang, Xingyu Luo, and Zhou Nie

Abstract

RNA virus infection is a global health issue with a significant economic burden. Direct visualization of the viral RNA genome in living cells is crucial for virological research and early clinical diagnosis. Thus, the need to continue research to find imaging toolkits is urgent. The RNA G-quadruplex (G4), a noncanonical secondary structure with stacked planar G-quartets, has recently been demonstrated in the RNA genomes of various viruses, especially the hepatitis C virus (HCV). Recent advancements in small molecular fluorescent probes have paved the way for a novel method of visualizing RNA G4s. Herein, we describe a fluorogenic probe-based RNA G4s light-up system for visualizing the HCV genome in living host cells and HCV RNA-presenting mini-organ-bearing mice without complicated sample pretreatment. Using this approach, we achieved (i) the visualization of HCV RNA genome in living cells, (ii) the investigation of viral RNA subcellular distribution, (iii) the dynamic tracking of native HCV infection and propagation in the host cell, and (iv) the high-contrast HCV RNA imaging in living mice bearing the HCV RNA-presenting mini-organ.

Key words Hepatitis C virus, RNA genome, G4, Fluorescence probes, In vivo imaging

1 Introduction

Chronic hepatitis C (CHC) is a global health issue with a significant economic burden resulting in liver failure, cirrhosis, and hepatocellular carcinoma [1, 2]. The World Health Organization (WHO) estimates that over 100 million people suffer from chronic HCV infection [3]. It is of great significance to realize the optical detection of native viruses for virological research as well as infection diagnosis. Visualization of viruses in living cells would be essential for unveiling the spatiotemporal dynamics of viral RNA translation, replication, and localization. The approaches for virus imaging can also be exploited to evaluate infection progression and treatment efficacy [4]. Recently, chemical labeling [5] has been used to visualize viruses inside living cells. However, this approach is limited by the low throughput, the difficulty in preparing these RNAs, and the

Shin-ichi Sato (ed.), *Live-Cell RNA Imaging: Methods and Protocols*, Methods in Molecular Biology, vol. 2875,
https://doi.org/10.1007/978-1-0716-4248-1_11,

complex procedure of microinjection. Other methods include molecular beacons and genetic modification of the viral genome [6]. Unfortunately, these approaches require a custom design for each viral genome sequence, and the transfected sequence can exhibit nonspecific nuclear sequestration. These issues prevented adoption by investigating naturally occurring viruses in clinical samples.

RNA G-quadruplex is a noncanonical secondary structure with stacked planar G-quartets formed by guanine-rich RNA sequences via Hoogsteen hydrogen bonding [7]. Recent studies have described the presence of RNA G4s in virus genomes, including HIV-1 (human immunodeficiency virus) [8], ZIKV (Zika virus) [9], SARS Co-V (severe acute respiratory syndrome coronavirus) [10], EBOV (Ebola virus) [11], MARV (Marburg virus) [11], and HCV [12]. These G4 structures are highly conserved in RNA virus genomes, making them promising targets for their detection [13]. Inspired by these facts, we envisage that the unique noncanonical structure of G4s in RNA viral genomes might be a potential target for developing a novel fluorogenic probe for viral RNA genome visualization. In the beginning, we performed the sequence alignments of the core gene of HCV, the most highly conserved structural protein in the HCV genome, and the regions rich in the putative G4 sequence [12]. In our research, we investigated the HCV subtype 2a, which is one of the most epidemic genotypes in China. The bioinformatics analysis results from 255 HCV subtype 2a-infected patients' HCV genome sequences (*see* **Note 1**) show a highly conserved G-rich consensus sequence (5′- GGGAAUGAGGGACUCGGCUGGGCAG GAUGG-3′) in the core gene between positions +259 and + 288, named CG2a (Scheme 1).

In these regards, we introduced two RNA G4-targeting lighting-up fluorogenic probes, **ThT-NE** [14] and **NIR-2** [15], to visualize HCV genomes in the living cell and mice. **ThT-NE** contains a **ThT** scaffold, a benzothiazole-based G4-specific fluorogenic probe extensively used for in vitro G4 analysis [16]. After binding with G4s, **ThT-NE** turns on its brightly green emission at 495 nm. **NIR-2** was rationally derived from red fluorescent protein chromophores and obtained infrared-shifted spectra at 689 nm, which are suitable for in vivo imaging [17] (*see* **Note 2**). After binding to HCV genome G4, both probes show a significant fluorescent enhancement, enabling lighting-up viral G4s. Using these probes, we describe the protocol of (i) characterization of the specificity of probe toward HCV RNA genome in vitro; (ii) cellular imaging of HCV RNA G4 mimicking system; (iii) imaging HCV genome in live host cells; (iv) verification of the specific targeting of probes toward the HCV core sequence and its innate G4 structure in cells; (v) monitoring the subcellular distribution of viral RNA genome; (vi) live-cell tracking of viral

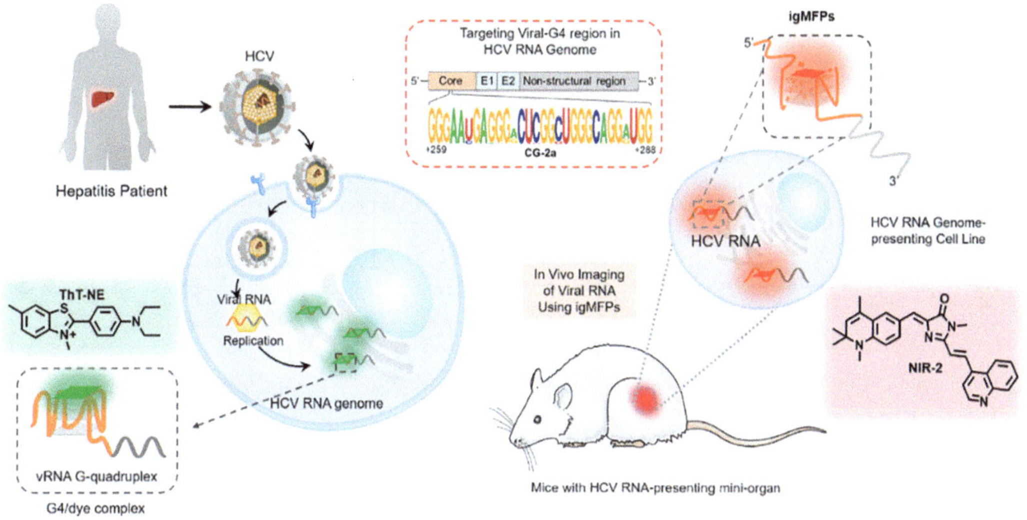

Scheme 1 Left: visualization of native hepatitis C virus RNA in living cells by **ThT-NE**. Right: visualization of full-length hepatitis C virus genome in mice by **NIR-2**

infection and propagation of clinically isolated HCV; and (vii) visualization of HCV RNA genome in GG2 cell mini-organ-bearing mice (*see* **Notes 3** and **4**). This state-of-art method is sample pretreatment-free and can directly visualize native viral genomes (Fig. 1).

2 Materials

1. Imaging probes: HCV RNA genome probes **ThT-NE** and **NIR-2** were synthesized following the reported procedures [14, 15]. (*See* **Note 5**).
2. Fluorescence measurement experiment: Different oligonucleotides (CG2a, GGG AAU GAG GGA CUC GGC UGG GCA GGA UGG; CG2a-M, GAG AAU GAG AGA CUC AGC UGA GCA AGA UAG; and CG2a-M2, GAA AAU GAG AGA CUC AGC UAA GCA AGA UAG) are purchased and purified using the polyacrylamide gel electrophoresis (PAGE) method; buffer, KCl (100 mM), Tris (25 mM, pH = 7.8), and DEPC water.
3. Cell culture: Hepatocyte-derived cellular carcinoma Huh7 and Huh7.5 cells were kindly provided by Charles Rice (Rockefeller University, New York); FCA1 cells, cells derived from Huh7 cells expressing the HCV subgenome without the core gene sequence; GG2 cells, cells expressing the HCV 2a genome; Dulbecco's Modified Eagle's Medium (DMEM) supplemented with 10% fetal bovine serum.

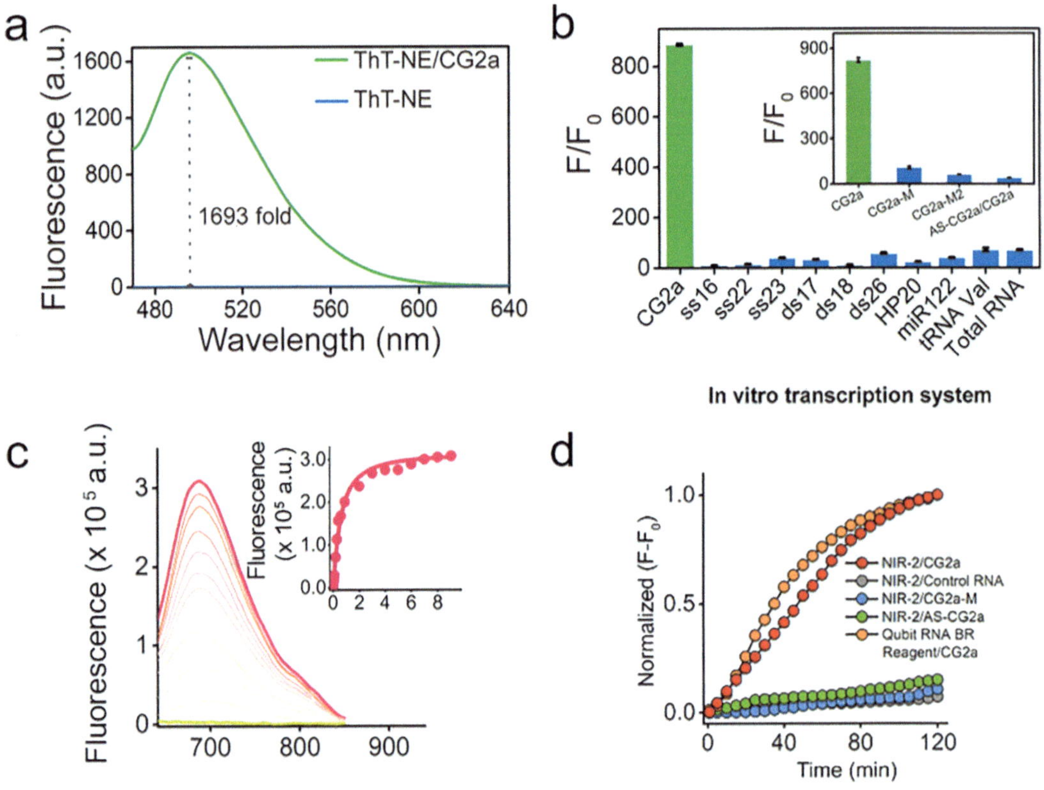

Fig. 1 In vitro characterization of the HCV RNA genome-specific fluorogenic function of probes. (**a**) Fluorescence spectra of **ThT-NE** (1 μM) in the absence or presence of CG2a (10 μM). (**b**) Changes in the fluorescence of **ThT-NE** (1 μM) in response to CG2a (5 μM) and other RNA conformations. The inset plots the fluorescence enhancement of **ThT-NE** (1 μM) in response to CG2a (5 μM) and its mutated sequences. (**c**) Emission spectra and fluorescence intensity change (insert) at 689 nm of **NIR-2** (1 μM) upon stepwise addition of G4 c-myc. d) Fluorescence kinetics of in vitro transcription of control RNA, CG2a-Mut, CG2a, and CG2a with AS-CG2a in the presence of **NIR-2** (10 μM) and CG2a in the presence of the Qubit RNA BR reagent (10×)

4. Cell imaging experiment, 35 mm confocal dish; staining of live cells, stock solution of Hoechst33342 and Golgi-Tracker, ER Tracker, MitoTracker, and LysoTracker; immunofluorescence, stock solution of DAPI; plasmid pSANG10-3F-BG4; anti-FLAG antibody; rhodamine-conjugated secondary antibody (goat anti-mouse IgG(H&L)-HRP); Fiji-Image J.
5. In vivo imaging and tissue staining: NCG (NOD-Prkdcem26Il2rgem26Nju) mice; staining of issue, OCT, 30% sucrose, 5% BSA/TBST. FAM-Beacon DNA (FAM- CGTTGAGAGGGGCGAGAGCCTCGGGGGA CATCAACG -BHQ-2) is purchased and purified using the PAGE method.

3 Methods

3.1 Characterization of the Fluorescent Probe to Label HCV Core Gene In Vitro

The HCV RNA genome selectively turn-on ability of probes **ThT-NE** and **NIR-2** is characterized in vitro (Fig. 1).

3.1.1 Fluorescence Measurement In Vitro

Prepare the stock solutions (100 μM) of the oligonucleotides by dissolving oligonucleotides directly into their respective buffers (*see* **Note 6**). Heat oligonucleotide solution at 65 °C for 5 min, and cool slowly to room temperature, and then keep at 4 °C overnight (*see* **Note** 7). See the detail of fluorescent probe stock solution preparation in **Note 8**.

1.1 For **ThT-NE**: Select the core gene of HCV subtype 2a, CG-2a, and its mutated sequences (CG2a-M, CG2a-M2, 5 μM) and the hybrid of CG2a and AS-CG2a (5 μM) (*see* **Note 9**), single-stranded (ss16, ss22, ss23, 15 μM), double-stranded (ds17, ds18, ds26, 15 μM), and hairpin (HP20, 15 μM) RNA, micro-RNA mimics (miR122, 15 μM), tRNA fragments (tRNA-Val, 15 μM), or total RNA of Huh7 cells (*see* **Note 10**).

1.2. For **NIR-2**: Select parallel G4 (c-myc and 27myc), mix-type G4 (H-telo, HT, and Hum21), antiparallel (TBA), single-stranded (SSAF22), double-stranded (ds26), and hairpin (HP18) (0–9 μM).

2. Mix 1 μL fluorescent probe (100 μM, in DMSO), 5 μL oligonucleotides (100 μM, in oligonucleotide buffer), and 84 μL incubation buffer (100 mM KCl, pH = 7.8, 25 mM Tris) in a 200 μL EP tube, affording a final total volume of around 100 μL.
3. Incubate the mixture for 15 min at room temperature.
4. Acquire fluorescence spectra using a QuantaMaster™ fluorescence spectrophotometer (PTI, Canada), and use the quartz cuvette with 2 mm × 10 mm path length for spectrum measurement optical path length of 10 mm. Keep both the excitation and emission slit widths at 5 nm.
5. Calculate the fluorescence enhancement by F/F_0. Herein, F is the fluorescence intensity of the probe and oligonucleotide complex, and F_0 is the fluorescence intensity of the probe itself.

3.1.2 Real-Time Monitoring of the RNA Transcript of CG2a Generated by T7 RNAP In Vitro

In this experiment, **NIR-2** is used to test the RNA transcript of CG2a generated by the in vitro T7 RNA polymerase (T7 RNAP) transcription system to mimic the RNA-generating mechanism in cellular conditions.

Prepare the template DNA solutions (10 μM, *see* **Note 11**) of the oligonucleotides by dissolving oligonucleotides directly into their respective buffers (*see* **Note 6**). Heat oligonucleotide solution at 65 °C for 5 min, and cool slowly to room temperature, and then keep at 4 °C overnight (*see* **Note 7**). See the detail of fluorescent probe stock solutions' preparation in **Note 8**.

1. Mix 2 μL template DNA (10 μM), 15.1 μL DEPC water (25 mM), 1.2 μL NTPs (25 mM), 2.7 μL $MgCl_2$ (9 mM), 3 μL Qubit RNA BR Reagent (10×), 1 μL **NIR-2** (300 μM), and 1 μL T7 RNA polymerase (200 U/μL) in a 200 μL EP tube, affording a final total volume of around 30 μL.
2. Monitor transcription at 37 °C for 120 min. During this time, record the fluorescence intensity every 1 min using Synergy™ Mx multi-mode microplate reader.

3.2 Cellular Imaging of HCV RNA G4 Mimicking System (Fig. 2)

Seed Huh7 cells at 35 mm confocal dish in a DMEM with 10% FBS at 37 °C in a 5% CO_2 atmosphere. Huh7 cells are a hepatocyte-derived cellular carcinoma cell line that has been widely used as a host for HCV infection [18].

1. Seed Huh7 cells (1×10^5) in a 35 mm confocal dish followed by incubation for 24 h.
2. Transfect fluorescence (FAM or Cy5)-tagged RNA (800 nM in 100 μL DMEM) using Lipofectamine™ 3000 transfection reagent (4 μL) following the standard protocol.
3. Incubate the cells for 1 h in a 37 °C incubator.
4. Wash cells with DEPC/PBS three times.
5. Incubate with **ThT-NE** or **NIR-2** (1 μM) for 10 min.
6. Wash the cells with DEPC/PBS three times. Acquire the fluorescent images using a Nikon confocal microscope (Nikon A1 plus). FAM and **ThT-NE** (green channel): $\lambda_{ex} = 488$ nm/$\lambda_{em} = 525 \pm 25$ nm. Cy5 and **NIR-2** (red channel): $\lambda_{ex} = 561$ nm/$\lambda_{em} = 700.5 \pm 37.5$ nm.

3.3 Visualization of the HCV Genome in Live Host Cells

Huh7 cells and HCV full-length genome express GG2 cells (*see* **Note 12**) are used to evaluate different dyes' imaging performance (Fig. 3).

1. Seed cells in a 35 mm confocal dish at a density of 1×10^5 cells per dish, and incubate with DMEM with 10% FBS at 37 °C in a 5% CO_2 atmosphere for 24 h for the following experiments.

2.1 For the comparison of **ThT-NE** and **ThT**, wash with 1 × DEPC-PBS three times, and incubate the cells with **ThT-NE** (1 μM, containing 0.05% DMSO) for 5 or 10 min and **ThT** (1 μM, containing 0.05% DMSO) for 30 or 240 min, respectively.

2.2. For the comparison of **NIR-2** and **ThT**, wash with 1 × DEPC-PBS three times, and incubate the cells with **NIR-2** (1 μM, containing 1% DMSO) or **ThT** (1 μM, containing 0.05% DMSO) for 10, 30, 60, and 120 min, respectively.

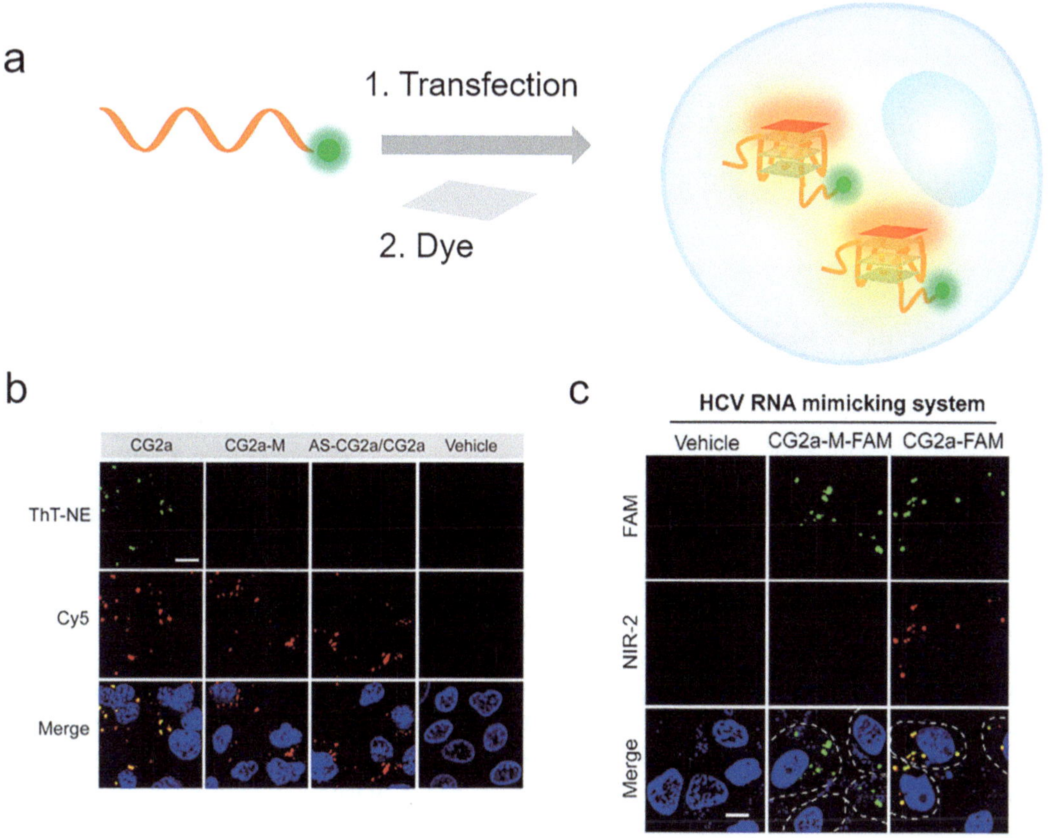

Fig. 2 Confocal imaging of transfected CG2a RNA at the live-cell level. (**a**) Schematic illustration of the HCV RNA mimicking system by transfecting hepatocytes (Huh7 cells) with CG2a RNA tagged with Cy5 or FAM dye to test the G4-specific imaging of dye. (**b**) Confocal imaging of living Huh7 cells transfected with various Cy5-tagged RNA sequences (red) such as CG2a, CG2a-M, and AS-CG2a/CG2a, respectively, and stained with 1 μM **ThT-NE** (green). Scale bar: 10 μm. (**c**) Confocal microscopy fluorescent imaging of the 1 μM **NIR-2**-stained Huh7 cells transfected with FAM-tagged CG2a RNA and FAM-tagged CG2a-M RNA, respectively. Scale bar: 10 μm. **ThT-NE** and FAM: $\lambda_{ex} = 488$ nm/$\lambda_{em} = 525 \pm 25$ nm. **NIR-2** and Cy5: $\lambda_{ex} = 561$ nm/$\lambda_{em} = 700.5 \pm 37.5$ nm. Hoechst33324: $\lambda_{ex} = 405$ nm/$\lambda_{em} = 445 \pm 25$ nm

3. Wash cells with 1 × DEPC/PBS three times.
4. Stain the nuclei using Hoechst33324 (blue) for 30 min.
5. After incubation, wash cells with 1 × DEPC/PBS three times.
6. Record the fluorescent images by using a Nikon confocal microscope (Nikon A1 plus). For **ThT-NE** and **ThT**, use green channel: $\lambda_{ex} = 488$ nm/$\lambda_{em} = 525 \pm 25$ nm. For **NIR-2**, use near-infrared channel: $\lambda_{ex} = 561$ nm/$\lambda_{em} = 700.5 \pm 37.5$ nm. Hoechst33324: $\lambda_{ex} = 405$ nm/$\lambda_{em} = 445 \pm 25$ nm.

In this experiment, the specificity of **ThT-NE** and **NIR-2** for the HCV viral RNA (vRNA) genome is evaluated by performing live-cell imaging analysis using both the probes and an HCV vRNA-

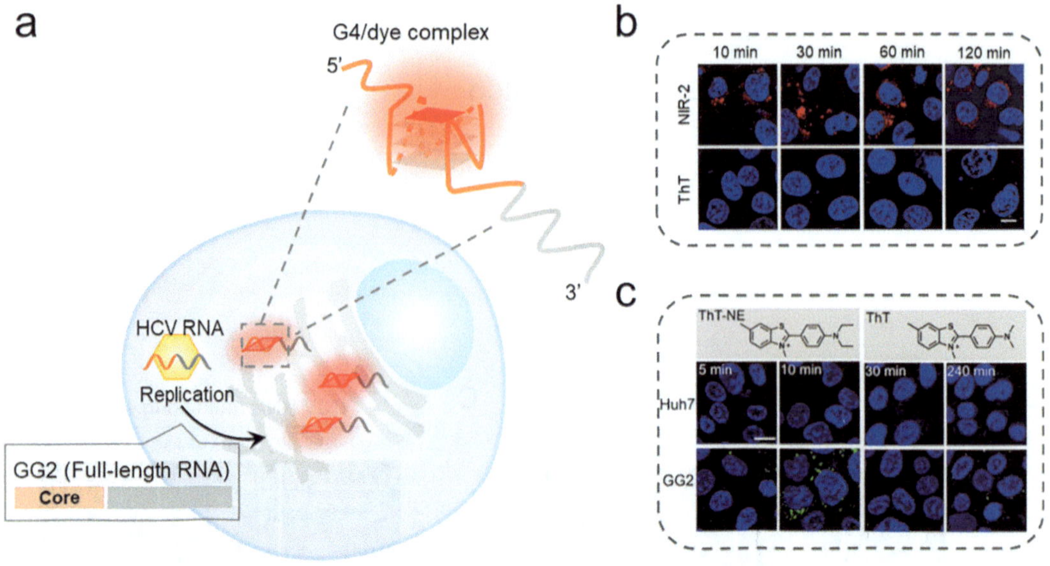

Fig. 3 Cellular imaging of the full-length HCV genome in living host cells. (**a**) Schematic illustration of fluorescence probes for tracking HCV RNA in the HCV full-length genome-presenting cell line GG2. (**b**) Confocal images of living GG2 cells at the indicated incubation times in the presence of **NIR-2** (1 μM) or **ThT** (1 μM). Scale bar: 10 μm. c) Representative confocal images of living Huh7 cells and GG2 cells at the indicated incubation times in the presence of **ThT-NE** (1 μM) or **ThT** (1 μM). Scale bar: 10 μm. **NIR-2**: $\lambda_{ex} = 561$ nm/$\lambda_{em} = 700.5 \pm 37.5$ nm. **ThT**: $\lambda_{ex} = 488$ nm/$\lambda_{em} = 525 \pm 25$ nm. **ThT-NE**: $\lambda_{ex} = 488$ nm/$\lambda_{em} = 525 \pm 25$ nm. Hoechst33324: $\lambda_{ex} = 405$ nm/$\lambda_{em} = 445 \pm 25$ nm

3.4 Evaluation of the Specificity of the Probes for the HCV Viral RNA Genome (Fig. 4)

specific molecular beacon (MB) probe. The MB probe with the Cy5/BHQ3 (fluorophore/quencher) pair (MB-C) is designed for hybridization with and turn-on detection of the conserved sequence (the position from +293 to +317) of the core gene in HCV genomic RNA.

3.4.1 Transfection of the Molecular Beacon (MB) Targeting the HCV Viral RNA

1. Seed the GG2 cells (vRNA+/core+/G4+), Huh7 cells (vRNA−/core−/G4−), FCA1 cells (vRNA+/core−/G4−) (20), and CG2a-mutated GG2-G4-Mut cells (about 1×10^5) (12) in a 35 mm confocal dish at 37 °C in a 5% CO_2 atmosphere for 24 h (*see* **Note 13**).

2.1 Transfect the MB probe with the Cy5 or FAM /BHQ3 (fluorophore/quencher) pair (MB-C) (400 nM in 100 μL DMEM) using Lipofectamine™ 3000 transfection reagent (4 μL) (*see* **Note 14**).

2.2 Transfect the AS-CG2a-hairpin (2 μM) in GG2 cells using Lipofectamine™ 3000 transfection reagent (4 μL) (*see* **Note 14**).

3. Incubate the cells for 1 h in a 37 °C incubator.
4. Wash the cells with DEPC/PBS three times.
5. Incubate with **ThT-NE** or **NIR-2** (1 μM) for 10 min.

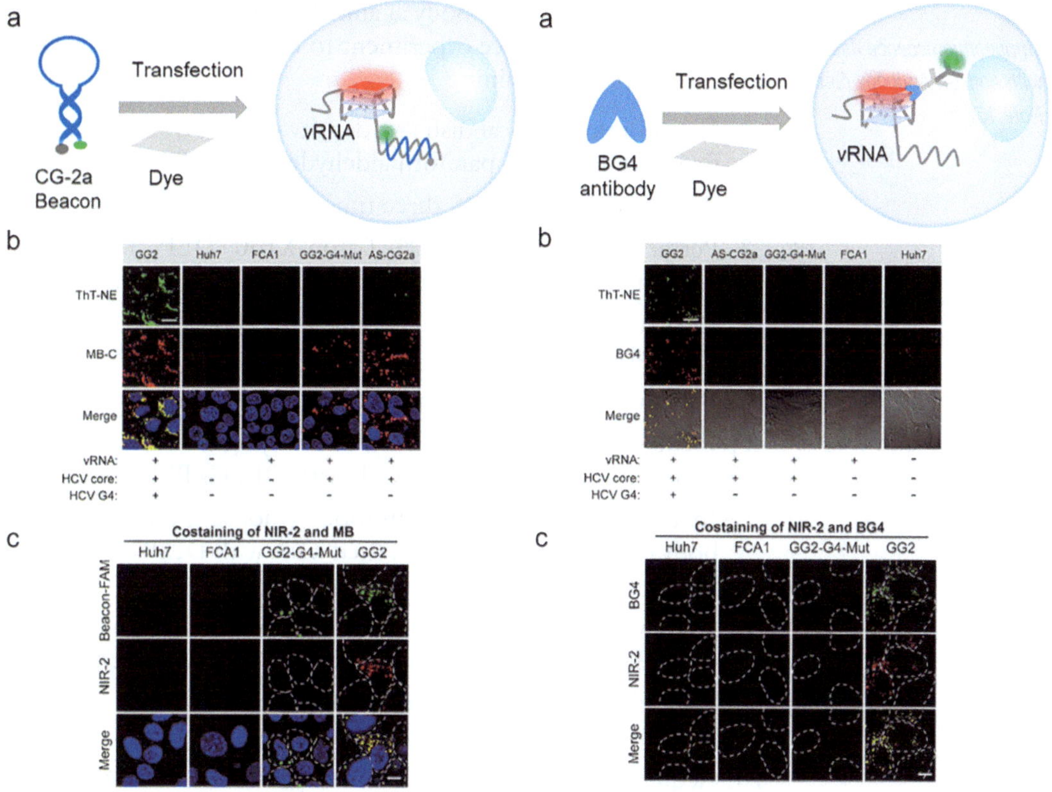

Fig. 4 Verification of the specific targeting of probes toward the HCV core sequence and its innate G4 structure in cells. (**a**) Schematic illustration of the HCV core gene targeted by dye and MB-C, an HCV core sequence-specific molecular beacon probe. (**b**) Confocal imaging of MB-C-transfected living host cells, including GG2 cells (vRNA+/core+/G4+), Huh7 cells (vRNA−/core−/G4−), FCA1 cells (vRNA+/core−/G4−), CG2a-mutated GG2-G4-Mut cells (vRNA+/core+/G4−), and AS-CG2a-transfected GG2 cells (vRNA+/core+/G4−) in the presence of **ThT-NE** (1 μM). Scale bar: 10 μm. (**c**) Confocal imaging of Beacon-FAM-transfected living Huh7, FCA1, GG2-G4-Mut, and GG2 cells in the presence of **NIR-2** (1 μM). Scale bar: 10 μm. (**d**) Schematic illustration of colocalization between dye and immunofluorescence staining of G4 structure on the HCV core gene by a BG4 antibody in cells. (**e**) Immunofluorescence staining of BG4 antibody and **ThT-NE** (1 μM) staining in the fixed GG2 cell without or with transfection of AS-CG2a, GG2-G4-Mut cell, FCA1 cell, and Huh7 cell. (**f**) Confocal images of the immunofluorescence staining of BG4 antibody and 1 μM **NIR-2** staining in the fixed Huh7, FCA1, GG2-G4-Mut, and GG2 cells. Scale bar: 10 μm. **NIR-2** and Cy5: $\lambda_{ex} = 561$ nm/$\lambda_{em} = 700.5 \pm 37.5$ nm. **ThT-NE** and FAM: $\lambda_{ex} = 488$ nm/$\lambda_{em} = 525 \pm 25$ nm. Hoechst33324: $\lambda_{ex} = 405$ nm/$\lambda_{em} = 445 \pm 25$ nm

6. Wash with DEPC/PBS three times.
7. Record the fluorescent image results using a Nikon confocal microscope (Nikon A1 plus). FAM and **ThT-NE** (green channel): $\lambda_{ex} = 488$ nm/$\lambda_{em} = 525 \pm 25$ nm. Cy5 and **NIR-2** (red channel): $\lambda_{ex} = 561$ nm/$\lambda_{em} = 700.5 \pm 37.5$ nm.

3.4.2 Immunofluorescence Assay with Anti-G4 Antibody (See Note 15)

BG4, a highly specific antibody against G4 structure, is exploited in this immunofluorescence experiment to compare their localization with probe-positive staining.

1. In a 35 mm confocal dish at a density of 1×10^5 cells per dish, fix the cell with 4% paraformaldehyde at 4 °C for 20 min.
2. Wash with DEPC/PBS three times.
3. Permeabilize cells with 0.2% Triton-X100/DEPC/PBS in for 12 min, and then wash with DEPC/PBS three times.
4. Incubate the permeabilized sample with 200 U/mL DNase I at 37 °C for 4 h.
5. Block with 20% goat serum at 37 °C for 1 h.
6. Incubate cells with 1 μM BG4 antibody against G-quadruplex at 37 °C for 1 h, and then wash with DEPC/PBS three times.
7. Incubate with anti-FLAG antibody (mouse monoclonal antibody) at 4 °C overnight, and then wash with DEPC/PBS three times.
8. Stain with fluorescence-conjugated secondary antibody at 37 °C for 30 min, and then wash with DEPC/PBS three times.
9. Stain cells with **ThT-NE** or **NIR-2** (1 μM) for 10 min, and then wash with DEPC/PBS three times.
10. Acquire the fluorescent images using a Nikon confocal microscope (Nikon A1 plus). For **ThT-NE**, use green channel: $\lambda_{ex} = 488$ nm/$\lambda_{em} = 525 \pm 25$ nm. For **NIR-2**, use near-infrared channel: $\lambda_{ex} = 561$ nm/$\lambda_{em} = 700.5 \pm 37.5$ nm. Hoechst33324: $\lambda_{ex} = 405$ nm/$\lambda_{em} = 445 \pm 25$ nm.

3.5 Monitoring the Subcellular Distribution of Viral RNA Genome (Fig. 5)

1. Seed about 10^5 cells at a 35 mm confocal dish in a DMEM with 10% FBS at 37 °C in a 5% CO_2 atmosphere for 24 h for the subsequent experiments. For staining of multiple host cellular organelles, incubate the GG2 cells with ER Tracker red (endoplasmic reticulum probe, 500 nM for 30 min), LysoTracker red (lysosome probe, 1 μM for 20 min), MitoTracker red (mitochondrion probe, 1 μM for 30 min), and Golgi-Tracker red (Golgi apparatus probe, 1 μM for 30 min) to label the subcellular organelles, respectively.
2. Wash the cells with 1 × DEPC-PBS three times.
3. Stain the cells with **ThT-NE** (1 μM, containing 0.05% DMSO) or **NIR-2** (1 μM, with 1% DMSO) for 10 min.
4. For the actin filaments staining, fix the GG2 cells with 4% paraformaldehyde for 25 min.
5. Incubate the fixed cells with TRITC-Phalloidin.
6. Wash with DEPC-PBS.

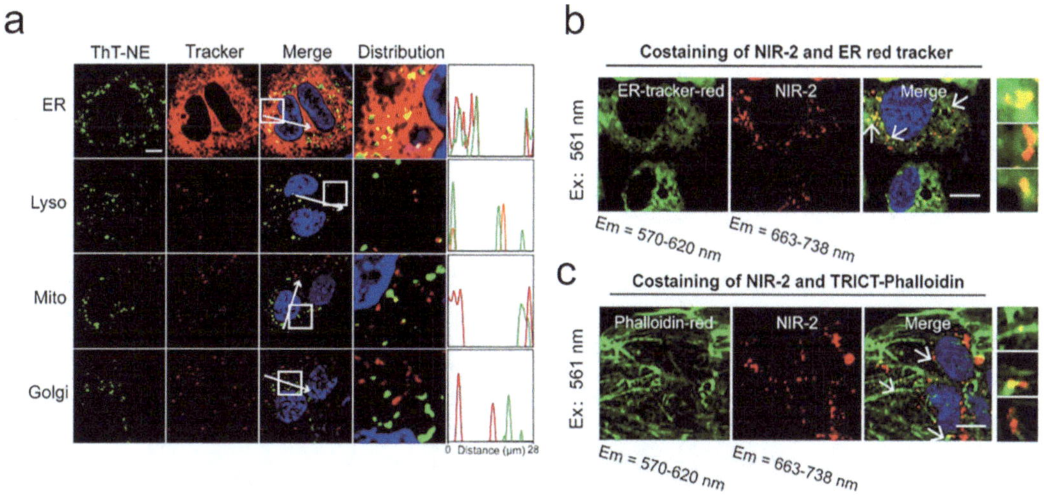

Fig. 5 Monitoring the subcellular distribution of the viral RNA genome. (**a**) Confocal imaging of living GG2 cells costained with 1 μM **ThT-NE** (green) and different organelle trackers (red) for the endoplasmic reticulum (ER), lysosome (Lyso), mitochondria (Mito), and Golgi apparatus (Golgi). Scale bar: 10 μm. (**b**) Confocal imaging of GG2 cells stained with ER Tracker red in the presence of **NIR-2** (1 μM). (**c**) Confocal imaging of fixed GG2 cells stained with TRICT-Phalloidin in the presence of **NIR-2** (1 μM). **NIR-2**: $\lambda_{ex} = 561$ nm/$\lambda_{em} = 700.5 \pm 37.5$ nm. ER Tracker red or TRICT-Phalloidin: $\lambda_{ex} = 561$ nm/$\lambda_{em} = 595 \pm 25$ nm. Hoechst33324 and DAPI: $\lambda_{ex} = 405$ nm/$\lambda_{em} = 445 \pm 25$ nm

7. Incubate cells with **NIR-2** (1 μM, containing 1% DMSO) for 10 min.
8. Stain the nuclei using Hoechst33324 or DAPI (blue).
9. Record the fluorescent images using a Nikon confocal microscope (Nikon A1 plus). **ThT-NE** (green channel):$\lambda_{ex} = 488$ nm/$\lambda_{em} = 525 \pm 25$ nm. Golgi Tracker/ER Tracker (red channel): $\lambda_{ex} = 561$ nm/$\lambda_{em} = 595 \pm 25$ nm. MB-C/LysoTracker/MitoTracker (near-infrared channel): $\lambda_{ex/em} = 640/763$–$738$ nm. **NIR-2** (near-infrared channel): $\lambda_{ex} = 561$ nm/$\lambda_{em} = 700.5 \pm 37.5$ nm. Hoechst33324 and DAPI: $\lambda_{ex} = 405$ nm/$\lambda_{em} = 445 \pm 25$ nm.

3.6 Live-Cell Tracking of Viral Infection and Propagation of Clinically Isolated HCV (Fig. 6) (See Note 16)

1. Collect HCV sera (male, core+, virus titer equal to 1.11×10^7 IU/mL) in local hospital under the ethics approval. (*See* **Note 17**).
2. Incubate HLCZ01 cells by hepatitis C patient sera with preliminary infection (moi = 0.1) or healthy human serum for 12 h, and remove the sera, and then incubate by DMEM with 10% FBS for different days, respectively.
3. Seed about 10^5 native HCV infected HLCZ01 cells into the 35 mm confocal dish, and then incubate with DMEM with 10% FBS at 37 °C in a 5% CO_2 atmosphere for 24 h.
4. Wash with 1 × DEPC-PBS three times.

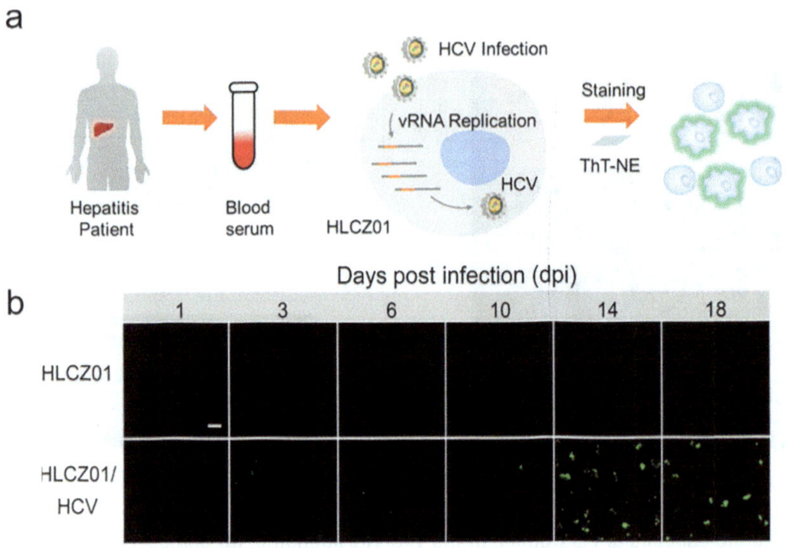

Fig. 6 Evaluation of native virus infection in living host cells. (**a**) Schematic illustration for dynamic monitoring of HLCZ01 cell infection by hepatitis C patient serum using **ThT-NE**. (**b**) High content imaging of **ThT-NE** (1 μM)-stained HLCZ01 cells infected with hepatitis C patient serum at 1–18 days post-infection (dpi). Scale bar: 100 μm. **ThT-NE**: $\lambda_{ex} = 488$ nm/$\lambda_{em} = 525 \pm 25$ nm

5. Incubate the cell in a DMEM with 10% FBS at 37 °C in a 5% CO_2 atmosphere for 24 h for the subsequent experiments.
6. Incubate the cells with **ThT-NE** (1 μM in DEPC-PBS) for 10 min at 1, 3, 6, 10, 14, and 18 days post hepatitis C patient serum infection.
7. Wash with 1 × DEPC-PBS three times.
8. Record the images with the Cytation 5 microplate reader using a full-field scanning model and green channel: $\lambda_{ex} = 488$ nm/ $\lambda_{em} = 525 \pm 25$ nm.

3.7 Visualization of HCV RNA Genome in GG2 Cell Mini-Organ-Bearing Mice (Fig. 7)

3.7.1 In Vivo Imaging

1. Inoculate the male NCG mice (5 weeks aged, 18–20 g) with Huh7 or GG2 mini-organ by injecting 10^6 cells in 200 μL DMEM subcutaneously near the leg (5 mice per group).
2. The mini-organ reaches 10 mm in diameter 21 days post cell injection.
3. Inject **NIR-2** (5 nmol, in 200 μL DEPC-PBS, with 0.5% DMSO) into the HCV RNA-presenting mini-organ and non-HCV RNA mini-organ.
4. Anesthetize the mice with isoflurane.
5. Record the real-time in vivo imaging at 5, 15, 30, 60, and 120 min after **NIR-2** injection using IVIS Lumina XR. **NIR-2**: $\lambda_{ex/em} = 640/695–770$ nm.

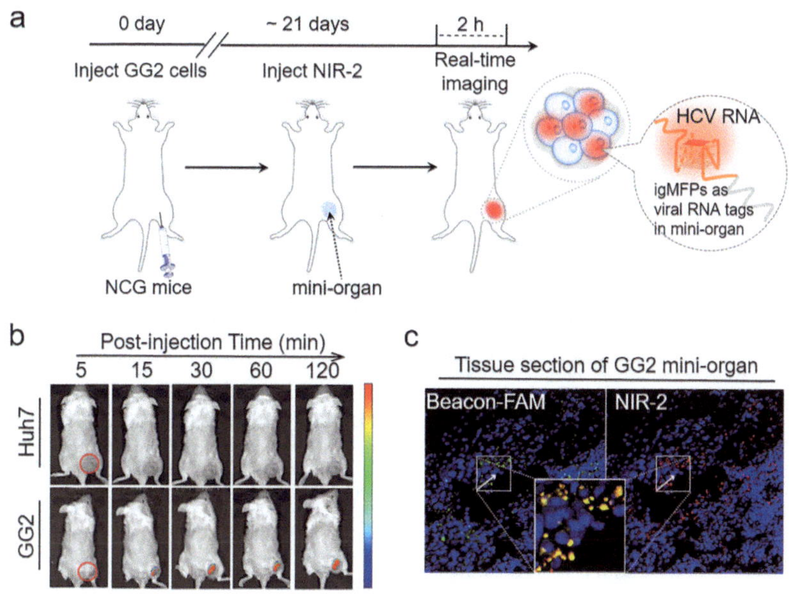

Fig. 7 (**a**) Schematic illustration of igMFPs as fluorescence tags for real-time near-infrared fluorescence (NIRF) visualization of HCV RNA in vivo. (**b**) Real-time NIRF visualization of the NCG mice after injection of **NIR-2** into the mini-organ embedded in the subcutaneous tissues. (**c**) Confocal fluorescence images of HCV RNA-presenting mini-organ tissue slices after the incubation with Beacon-FAM and **NIR-2**. (1 μM). Scale bar: 50 μm

3.7.2 Staining of Tissue Samples

1. After the imaging experiment, euthanize all the mice by cervical dislocation.
2. Resect the mini-organ.
3. Fix all tissue samples in 4% formaldehyde for 24 h at 4 °C.
4. Dehydrate in 30% sucrose until the samples sank to the bottom.
5. Embed the tissues with OCT in plastic molds, and freeze the OCT-containing tissues at −80 °C overnight.
6. Slice tissue samples into serial sections (20 μm in thickness) using a Leica CM1950 frozen section machine.
7. Treat the slices with 1 μM Beacon-FAM oligonucleotides (200 μL) for 2 h at 37 °C.
8. Wash the slices with DEPC-PBS three times.
9. Incubate cells with 1 μM **NIR-2** (200 μL DEPC-PBS, with 0.5% DMSO) for 30 min.
10. The nuclei are stained with DAPI.
11. Record the fluorescent images using a Nikon confocal microscope (Nikon A1 plus). DAPI (blue channel): $\lambda_{ex} = 405$ nm/$\lambda_{em} = 445 \pm 25$ nm. NS5A (green channel): $\lambda_{ex} = 488$ nm/$\lambda_{em} = 525 \pm 25$ nm. **NIR-2** (near-infrared channel): $\lambda_{ex} = 561$ nm/$\lambda_{em} = 700.5 \pm 37.5$ nm.

4 Notes

1. The core gene of HCV subtype 2a is obtained from the National Center for Biotechnology Information Website (www.ncbi.nlm.nih.gov) and the HCV database (www.hcv.lanl.gov/) and aligned using WebLogo software (https://weblogo.berkeley.edu/logo.cgi).
2. Photophysical properties of **ThT** [16, 19], **ThT-NE**, and **NIR-2** with G-quadruplex as shown in the following.

Probes	Molecular structures	$\lambda_{ex}/\lambda_{em}$ (nm)	Q.Y.	K_d (μM)	S/B
ThT[a]	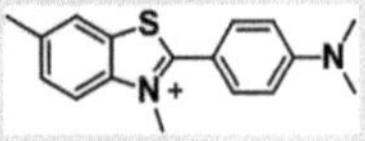	432/490	0.25	1.30 ± 0.15	1700
ThT-NE[b]	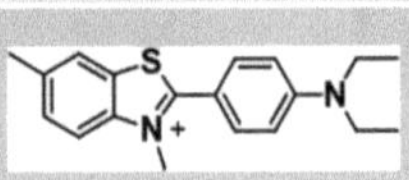	461/495	0.314	1.77 ± 0.19	1693
NIR-2[b]	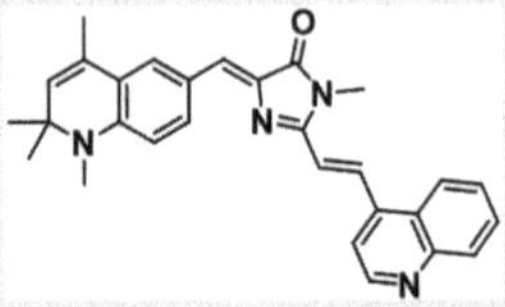	586/689	0.20	1.06 ± 0.05	78

[a] The photophysical properties of **ThT** were measured with parallel topological structure G-quadruplex 22AG; [b] the photophysical properties of **ThT-NE** and **NIR-2** are measured with parallel topological structure G-quadruplex CG2a; [c] the molar extinction coefficient ε of **ThT** alone.

3. In our work, **ThT-NE** and **NIR-2** show high specificity for HCV viral RNA G4 structure rather than endogenous G4s. The following reasons for this specificity are proposed:
   (a) The amount of the viral RNA G4s in infected cells (10^3–10^4 copies per cell) [20, 21] largely surpasses that of the host endogenous G4s (typically ~10^2 foci per cell observed by BG4 antibody staining and SiR-PyPDS staining) [22–24]. This explanation is supported by our previously fluorescence in situ hybridization (FISH) experiment, showing that the expression level of HCV viral RNA is considerably higher than those of other G4-containing RNAs (e.g., *NRAS*, *BCL-2*, *TERRA*, *VEGF*, *ZIC1*) in the cytoplasm (previously reported as Fig. S37 in the supplementary information of Ref. [14]). In addition, the unique replication machinery of the HCV genome increases the local concentrations of HCV viral

RNA, thus further improving the specificity of **ThT-NE** targeting HCV viral RNA G4. During HCV replication, the virus induces the remodeling of the host ER membrane to form vesicle-like viral replication complexes with an average diameter of about 150 nm [25]. According to previous findings on intracellular dynamics of HCV RNA synthesis [26], the virus infection induces remodeling of the host ER membrane to form vesicle-like viral replication complexes with approximately 10–50 copies of HCV RNA during HCV replication, further enlarging the local concentration difference between HCV G4 and endogenous G4. Furthermore, virus infection causes cell cycle arrest of host cells, leading to significantly decreased proportions of cells in the S phases [27]. However, most BG4 foci or SiR-PyPDS foci are observed during the S phase due to the replication-dependent formation of DNA G4 structures. Cell cycle arrest leads to a significant reduction in DNA G4 staining foci [23]. Based on these results, it is reasonable to suppose that the virus infection in host cells would inhibit the formation of nucleic DNA G4 structures, potentially leading to the reduced amount of endogenous G4s in host cells.

(b) The low concentration of the **ThT-NE** and **NIR-2** probes (1 μM) and short incubation time (10 min) enable negligible background fluorescence from endogenous G4s. The dynamic range characterization of the probe shows that **ThT-NE** and **NIR-2** with high probe concentration and long incubation time (3 μM **ThT-NE** incubation for 1 h or 10 μM **NIR-2** incubation for 2 h) could image endogenous G4s in Huh7 cells (**ThT-NE** located in the nucleoli and **NIR-2** located in the cytoplasm). However, these endogenous G4 signals in Huh7 cells are significantly weaker than those of viral G4 in GG2 cells.

(c) We propose a simplified theoretical model of the interaction between probes and G4 to elucidate the specific imaging of HCV RNA. As shown in Scheme 2, since the local concentration of HCV RNA G4 (B_{max2}) is significantly higher than that of host endogenous G4 (B_{max1}), the minimal intracellular concentration of probes required for host G4 to be detectable will be much higher than that for HCV G4 being detectable. The intracellular concentration of the probe is positively related to the concentration and incubation time for penetration, so the proper control of experimental settings will meet the requirement of an ultralow-background window.

4. Circular dichroism (CD) reveals that CG2a is folding into the parallel G-quadruplex topological structure with a positive

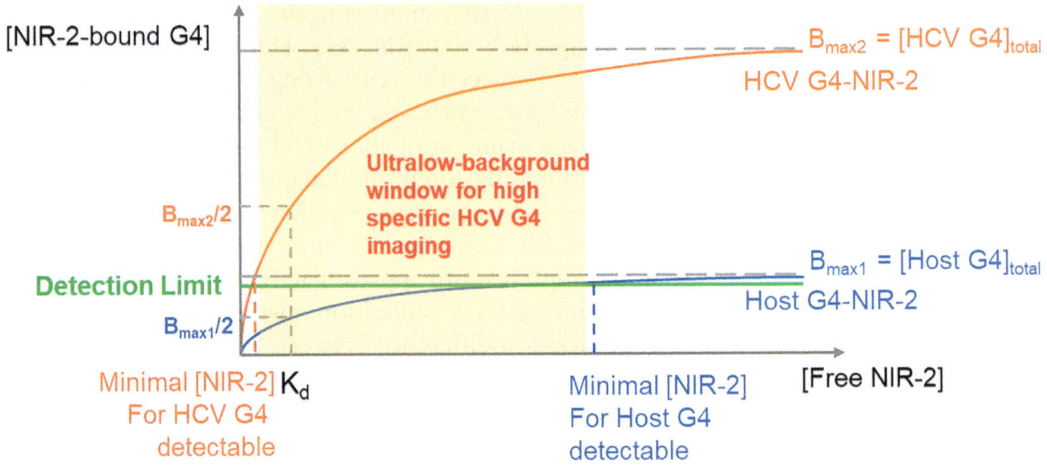

Scheme 2 Since the local concentration of HCV RNA G4 (B_{max2}) is significantly higher than that of host endogenous G4 (B_{max1}), the minimal intracellular concentration of probes required for host G4 to be detectable will be much higher than that for HCV G4 being detectable. The intracellular concentration of the probe is positively related to the concentration and incubation time for penetration, so the proper control of experimental settings will meet the requirement of an ultralow-background window

peak at 265 nm and a negative peak at 240 nm. There are no discernible alterations after incubation with our probe, indicating no significant difference in topological structure [28]. This analysis suggests that G4-specific **ThT-NE** and **NIR-2** possess unique merit that neither causes the global induction of G4 structures nor perturbs the endogenous G4 folding dynamics.

5. Due to the deliquescence of zinc chloride, it is necessary to minimize the exposure time of anhydrous zinc chloride in the air during weighing. Besides, the reaction equipment (flask, magnetic stirrer, and reflux tube) needs to be dried to reduce the moisture in the glass container.
6. All RNA oligonucleotides should be dissolved and completed with DEPC water in sterile culture. RNA oligo-related experiments should be dissolved and completed by DPBS in a sterile culture environment.
7. To form the G4 structure, all the oligonucleotides are annealed at 65 °C, then slowly cooled down to room temperature, and then stayed overnight at 4 °C in the buffer.
8. To avoid photobleaching and reduce the fluorescence quenching, the fluorogenic probes (powder) should be stored in the dark at −20 °C. The stock solutions are prepared by dissolving the fluorogenic probe in DMSO with a concentration of 2 mM. The working solutions are diluted by DEPC water into the corresponding concentration for relevant experiments.

9. Four guanines in CG2a (5′- GGG AAU GAG GGA CUC GGC UGG GCA GGA UGG-3′) are replaced by adenosines obtaining its mutated sequences CG2a-M (5′- G**A**G AAU GAG **A**GA CUC **A**GC UG**A** GCA **A**GA U**A**G-3′). This replacement will reduce Hoogsteen hydrogen bonding formation and block G-quadruplex folding. AS-CG2a is the antisense strand of CG2a.
10. Total RNA of Huh7 cells and GG2 is extracted from 3×10^5 cells using the TRIzol reagent.
11. The sequence of the template DNA coding strand with T7 promoter (underlined) used in this study is as follows:

Name	Sequence
CG2a template	5-TAATACGACTCACTATAGGG GGGAATGAGGGACTCGGCTGGGCAGGATGG-3'
CG2a-M template	5-TAATACGACTCACTATAGGG GAGAATGAGAGACTCAGCTAAGCAAGATAG-3'
AS-CG2a template	5-TAATACGACTCACTATAGGG CTCTTACTCTCTGAGTCGACTCGTTCTATC-3'
Control template	5-TAATACGACTCACTATAGGG TGGAGTGTGACAATGGTGTTTGT-3'

12. Huh7 is a hepatocyte-derived cellular carcinoma cell line widely investigated in HCV-relative research. GG2 is a full-length HCV 2a infectious cell line derived from the hepatoma cell line Huh7 and expresses the full-length genome of HCV subtype 2a (JFH-1) [27, 29].
13. FCA1 cells are derived from Huh7 cells transfected with HCV subgenomic replicons I377/NS3–3′ (Con1) [30]. GG2-G4-Mut cells express the HCV 2a genome with multiple point mutations in the G4-forming CG2a region (+265 to +279) of the core gene (GG2-G4-Mut cells was a gift from the State Key Laboratory of Virology and Hubei Province Key Laboratory of Allergy and Immunology and Department of Immunology, School of Medicine, Wuhan University) [12].
14. The MB probe with the Cy5 or FAM/BHQ3 (fluorophore/quencher) pair (MB-C) is designed for hybridization with and turn-on detection of the conserved sequence (the position from +293 to +317) of the core gene in HCV genomic RNA. AS-CG2a-hairpin is a sequence entirely complementary to CG2a. AS-CG2a-hairpin (2 μM) is transfected in GG2 cells for G4 blocking.
15. Here, the indirect immunofluorescence technique is used. BG4, as a G4-specific antibody, is used to mark the G4 structure at first. Then the primary antibody, anti-FLAG (a mouse

monoclonal antibody), is used to bind BG4 specifically. Then the samples are incubated with fluorescence-conjugated Goat Anti-Mouse secondary antibody to combine with the primary antibody. The data is observed under a Nikon fluorescence microscope (Confocal Laser Scanning Microscope Nikon A1 plus (Japan)) [21].

16. All living viral experiments must be performed in a biosafety level 2 (BSL2) facility.
17. HCV sera are obtained from Xiangya Hospital of Central South University, Changsha, China, with the approval of the Ethics Committee of Xiangya Hospital.

References

1. Mohd Hanafiah K, Groeger J, Flaxman AD et al (2013) Global epidemiology of hepatitis C virus infection: new estimates of age-specific antibody to HCV seroprevalence. Hepatology 57:1333–1342
2. Jones CT, Catanese MT, Law LMJ et al (2010) Real-time imaging of hepatitis C virus infection using a fluorescent cell-based reporter system. Nat Biotechnol 28:167–171
3. Global hepatitis report, 2017. Geneva: World Health Organization; 2017. http://apps.who.int/iris/bitstream/10665/255016/1/9789241565455-eng.pdf?ua=1
4. Lindenbach BD, Rice CM (2005) Unravelling hepatitis C virus replication from genome to function. Nature 18:933–938
5. Santangelo PJ, Lifland AW, Curt P et al (2009) Single molecule–sensitive probes for imaging RNA in live cells. Nat Methods 6:347–349
6. Burch BD, Garrido C, Margolis DM (2017) Detection of human immunodeficiency virus RNAs in living cells using spinach RNA aptamers. Virus Res 15:141–146
7. Kwok CK, Merrick CJ (2017) G-Quadruplexes: prediction, characterization, and biological application. Trends Biotechnol 35:997–1013
8. Amrane S, Kerkour A, Bedrat A et al (2014) Topology of a DNA G-Quadruplex structure formed in the HIV-1 promoter: a potential target for anti-HIV drug development. J Am Chem Soc 136:5249–5252
9. Fleming AM, Ding Y, Alenko A et al (2016) Zika virus genomic RNA possesses conserved G-quadruplexes characteristic of the Flaviviridae family. ACS Infect Dis 2:674–681
10. Pratihar S, Agrawal R, Pal VKS et al (2022) Reliable fluorometric detection of SARS-CoV-2 by targeting the G-quadruplex through pH-triggered conformational polymorphism. ACS Sens 7:453–459
11. Krafcikova P, Demkovicova E, Viglasky V (2016) Ebola virus derived G-quadruplexes: thiazole orange interaction. Biochim Biophys Acta 1861:1321–1328
12. Wang SR, Min YQ, Wang JQ et al (2016) A highly conserved G-rich consensus sequence in hepatitis C virus Core gene represents a new anti–hepatitis C target. Sci Adv 2:e1501535
13. Ruggiero E, Richter SN (2018) G-quadruplexes and G-quadruplex ligands: targets and tools in antiviral therapy. Nucleic Acids Res 46:3270–3283
14. Luo X, Xue B, Feng G (2019) Lighting up the native viral RNA genome with a fluorogenic probe for the live-cell visualization of virus infection. J Am Chem Soc 141:5182–5191
15. Zhang J, Li H, Lin B et al (2021) Development of near-infrared nucleic acid mimics of fluorescent proteins for in vivo imaging of viral RNA with turn-on fluorescence. J Am Chem Soc 143:19317–19329
16. Mohanty J, Barooah N, Dhamodharan V et al (2013) Thioflavin T as an efficient inducer and selective fluorescent sensor for the human telomeric G-quadruplex DNA. J Am Chem Soc 135:367–376
17. Hoffman RM (2005) The multiple uses of fluorescent proteins to visualize cancer in vivo. Nat Rev Cancer 5:796–806
18. Ramakrishnaiah V, Thumann C, Fofana I et al (2013) Exosome-mediated transmission of hepatitis C virus between human hepatoma Huh7. 5 cells. Proc Natl Acad Sci USA 110: 13109–13113
19. Bukh J (2016) The history of hepatitis C virus (HCV): basic research reveals unique features in phylogeny, evolution and the viral life cycle

with new perspectives for epidemic control. J Hepatol 65:S2–S21

20. Zhang YY, Zhang BH, Ishii K et al (2010) Novel function of CD81 in controlling hepatitis C virus replication. J Virol 84:3396–3407
21. Biffi G, Tannahill D, McCafferty J et al (2013) Quantitative visualization of DNA G-quadruplex structures in human cells. Nat Chem 5:182–186
22. Biffi G, Di Antonio M, Tannahill D et al (2014) Visualization and selective chemical targeting of RNA G-quadruplex structures in the cytoplasm of human cells. Nat Chem 6:75–80
23. Di Antonio M, Ponjavic A, Radzevičius A et al (2020) Single-molecule visualization of DNA G-quadruplex formation in live cells. Nat Chem 12:832–837
24. Suzuki T (2017) Hepatitis C virus replication. Organelle Contact Sites 997:199–209
25. Vogt PK (2013) In: Bartenschlager R (ed) Hepatitis C virus: from molecular virology to antiviral therapy, vol 369. Springer, Berlin, pp 167–198
26. Walters KA, Syder AJ, Sl L et al (2009) Genomic analysis reveals a potential role for cell cycle perturbation in HCV-mediated apoptosis of cultured hepatocytes. PLoS Pathog 5: e1000269
27. Wakita T, Pietschmann T, Kato T et al (2005) Production of infectious hepatitis C virus in tissue culture from a cloned viral genome. Nat Med 11:791–796
28. Carvalho J, Queiroz JA, Cruz C (2017) Circular dichroism of G-quadruplex: a laboratory experiment for the study of topology and ligand binding. J Chem Educ 94:1547–1551
29. Qin Y, Xue BB, Liu CY et al (2017) NLRX1Mediates MAVS degradation to attenuate hepatitis C virus-induced innate immune response through PCBP2. J Virol 91:1–17
30. Hayashi J, Stoyanova R, Seeger C (2005) The transcriptome of HCV replicon expressing cell lines in the presence of alpha interferon. Virology 335:264–275

Chapter 12

Imaging of Viral Genomic RNA Replication with Nanoprobes

Ping Ping Hu, Lin Ling Zheng, Lei Zhan, and Cheng Zhi Huang

Abstract

Viruses are a great threat to human life and health. Different viruses have its unique mechanism to efficiently infect cells, and the entry process and the nucleic acid replication using cell machine are two critical processes related to the fate of virus progeny. Real-time and long-term imaging techniques can be used to thoroughly investigate the viral infection events. This chapter will present the labeling of viral genomic RNA (gRNA) replication by developing new nanoprobes, one-donor-two-acceptors one, for example, in which the synergistic coupling of multiple energy transfer strategy, so as that the journey of viruses in live cells could be monitored and imaged in real time. Methods of labeling as well as that used for fluorescent and dark field scattering imaging are outlined.

Key words Live-cell imaging, gRNA replication, Fluorescence resonance energy transfer, Nanoprobe

1 Introduction

Viruses constitute a class of the most infectious pathogens causative highest death toll in the world [1, 2]. Viral infectious diseases are responsible for millions of deaths and cause huge economic losses per year [3]. From late 2019, the new outbreak of coronavirus disease (COVID-19) caused by severe acute respiratory syndrome coronavirus 2 (SARS-CoV-2) has resulted in over 6 million deaths globally by May 13, 2022 [4], which seriously devoured the public health and lives and costed global economic development.

Although the pathogenic mechanism of each virus is diverse, the nucleic acid replication of all existing viruses using the host cells is critical related to the fate of virus progeny [5, 6]. Initial non-availability of effective clinical intervention of the virus disease highlighted the need of comprehensive understanding of viral infection mechanism. To achieve this goal, it's crucial to invest great endeavor on the precise tracking of pathogens for the mechanism of viral entry [7–9], intracellular trafficking and sorting [10, 11], as well as viral egress [12–14]. For that purpose, the viral genomic RNA (gRNA) labeled for long-term and real-time

Shin-ichi Sato (ed.), *Live-Cell RNA Imaging: Methods and Protocols*, Methods in Molecular Biology, vol. 2875, https://doi.org/10.1007/978-1-0716-4248-1_12,

imaging techniques is presented by incorporated nanoprobes, which shed light on revealing the events of viral gRNA replication and virus–host cell interaction.

Viruses are ubiquitous pathogens, which can use the energy and metabolic apparatus of host cells for replication and proliferation. Although the type of nucleic acids that the virus contains is various [15], all viruses have developed a series of sophisticated mechanisms for viral replication and pathogenetic events [5, 16]. Viral infection starts with binding to the specific receptor on the host cell membrane. Most viruses then release their genetic materials and transport them to appropriate organelle of the host cell, followed by the viral gRNA amplification and protein expression. Finally, the newly synthesized viral components assemble into a complete virion, which then buds from the cell. Thorough understanding of viral infection pathway not only has great significance for the active treatment and prophylaxis against viruses but also provides new insights into cell biology.

Over the past several years, single particle tracking (SPT) makes it possible to follow the journey of a virus particle or virus components in the host cells and to obtain detailed dynamic interaction between viruses and cell [5, 17–21]. The combination of SPT with time-resolved fluorescence imaging and living cell imaging technology and the movement of fluorescent labeled viruses or virus components in live cells is visualized, which is momentous for the accurate interpretation of virus infection mechanism. Moreover, with the development of nanotechnology, viral protein and/or viral gRNA is probed with various fluorescent nanoscale probes [8, 22–24], gold- or silver-based inorganic nanoparticles [25–27], and liposomes [28], paving the way for monitoring individual viral infection behaviors in live cells. However, exogenous probes always lack single molecule sensitivity, and most studies only focus on the location of virus in cells at rest or the response of cells to viral infection at a certain time of viral attachment, early cellular entry, or late egress. Very few insights are available about the life cycle of a virus and diverse routes of viral infection in real time and long term. Santangelo et al. developed a new method called multiple labeled tetravalent RNA imaging probes (MTRIPs) [29, 30]. In this work, the RNA/DNA oligonucleotides complementary to a given *virus'* genome were internally labeled with fluorophores and biotinylated on the 3′ and 5′ end, respectively. When mixed with neutravidin, these oligonucleotides were tetramerized, the brightness of which increased significantly, and achieved to monitor over the entire replication cycle of viral gRNA in live cells. However, this approach may also get a strong imaging background, and once the probes dissociate from the target nucleic acid, virus replication can no longer be visualized with this probes, resulting in false-positive signals.

Molecular beacon (MB), a successful example of nucleic acids based probes, is described by Tyagi and Kramer in 1996 firstly [31, 32]. MB is a kind of hairpin-shaped oligonucleotide probe, which consists of a loop that can bind to the target nucleic acids and a stem that can hybridize to form a double strand. MBs are usually modified with fluorescent dyes at one end and quenching groups at the opposite end. The unique stem–loop structure and fluorescence resonance energy transfer (FRET) between fluorophore–quencher pair are two key factors that determine MBs' functionality and efficiency [23, 33]. When there is no target, the loop of MBs is closed, and fluorophore and quencher in the two ends get into close proximity, resulting in the fluorescence effectively quenched. When there is a target DNA/RNA molecule, the loop sequences hybridize with the target, the stem helix of MBs opens, and the spatial separation between fluorophore and quencher facilitates fluorescence emitting. Due to the advantages of ultrahigh sensitivity, specificity, and signal-to-background ratio (S/N), MBs have been widely applied in exploring DNA–protein interaction and in real-time detection and imaging the key events during genetic material replication, transcription, and expression [26, 33, 34]. With the wide utilization of nanomaterials in various fields and the combination of nanotechnology with gene probing and visualization methods, inorganic fluorescent nanomaterials with higher fluorescence intensity, such as quantum dots (QDs), have been used as fluorescence donors to improve the fluorescence imaging sensitivity. The excellent optical stability of QDs also provides an opportunity to monitor the DNA/RNA replication process for a long time. To further improve the energy transfer efficiency and increase S/N, a series of super-quenching systems are developed, by increasing the number of quenchers [35, 36] or using some nanomaterials (e.g., AuNPs [37], graphite nanoparticles [38], etc.) as energy transfer receptors. Among all of the MB-based platforms mentioned above, QDs–AuNPs pair are attractive and widely applied in protease analysis [39] and virus detection and tracking [40, 41], due to their easy preparation and handling and high efficiency of energy transfer. To avoid high false-positive signal of QDs unbound or dissected from the targets, effective quenching of QDs fluorescence requires the approach of multiple AuNPs at the same time, which makes it difficult to realize the one-to-one signal output of a single target molecule and a single quantum dot, leading to reduced imaging sensitivity.

To improve the energy transfer efficiency of QDs–AuNPs pairs, we designed a novel type of nanoprobes (one-donor-two-acceptors, DAA*) using dual energy transfer at the same time to monitor the viral gRNA replication in long term and real time [26]. As showed in Fig. 1, the DAA* nanoprobe was constructed with streptavidin conjugated QDs (SA-QDs) as the fluorescent donor, while AuNPs (A) and multiple BHQ-2 (A*, a commonly used

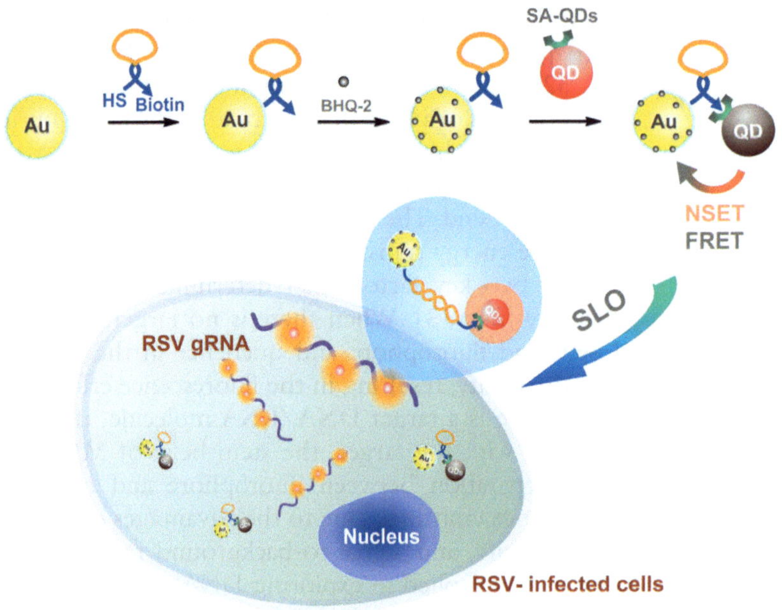

Fig. 1 Schematic diagram of one-donor-two-acceptors (DAA*) nanoprobe for viral gRNA replication in live cells based on QDs and AuNPs

quencher for organic dyes) were combined to form a composite acceptor (AA*). Taking respiratory syncytial virus (RSV) as an example, a MB molecule was involved as the connecter between energy donor and receptor, the loop of which is complementary to a conserved and repeated sequence of RSV gRNA. The two ends of MBs were modified with biotin and thiol, respectively, which can interact with SA-QDs and BHQ-2-loaded AuNPs. DAA* probes were successfully assembled through the ultrahigh affinity between biotin and streptavidin, as well as the strong Au–S covalent linkage. The exciton–plasmon interaction between QDs and AuNPs (NSET) and exciton–dipole interaction between QDs and BHQ-2 (FRET) can therefore coupling occur, which demonstrated a high energy transfer efficiency and low S/B for real-time and long-term imaging of viral gRNA replication. Additionally, AuNPs are the most commonly used nanoprobes in dark field scattering imaging (iDFM), which is also the main components of the DAA* nanoprobe. In addition to fluorescence imaging, these composite probes can also be used for iDFM of viral gRNA replication in live cells. Collectively, the QDs–AuMBQ probes with outstanding photostability can be easily prepared and repeated delivered into multiple cell types and achieved to follow the lives of most viruses, which may become the standard probes for future viral infection studies.

2 Materials

1. Bis(p–sulfonatophenyl) phenylphosphine dipotassium salt (BSPP).
2. Gold nanoparticles (AuNPs).
3. MWCO 30 KD Nanosep device.
4. Biotin–MB–SH (*see* **Note 1**).
5. Blocking DNA (*see* **Note 1**).
6. Tris(2–carboxyethyl) phosphine hydrochloride (TCEP).
7. Sodium citrate–HCl buffer solution (pH 3.0).
8. BHQ–2 amine.
9. SA–QDs solution (*see* **Note 2**).
10. Zonyl® FSN fluorosurfactant.
11. Streptolysin O.
12. Cysteine solution.
13. Paraformaldehyde.
14. Bovine serum albumin (BSA).
15. Silver enhancer kit.
16. Sodium thiosulfate solution.

3 Methods

3.1 Preparation of AuNPs–MBs Conjugates

1. Add 200 μL 10 mM bis(p-sulfonatophenyl) phenylphosphine dipotassium salt (BSPP) to 1.0 mL 83 nM 5 nm AuNPs solution overnight to block the nonspecific adsorption of oligonucleotides on the surface of AuNPs. Remove the excess BSPP via centrifugal ultrafiltration using MWCO 30 KD Nanosep device.
2. Activate Biotin–MB–SH or blocking DNA with Tris(2–carboxyethyl) phosphine hydrochloride (TCEP) for 1 h at a molar ratio of 1:50.
3. Mix 0.5 mL BSPP stabilized AuNPs solution (Au–BSPP, 83 nM) with 0.5 mL active Biotin–MB–SH (83 nM) and blocking DNA (415 nM) mixed solution.
4. Add 42 μL 500 mM sodium citrate–HCl buffer solution (pH 3.0) to above mixture.
5. Incubate the mixture for 10 min at room temperature.
6. Transfer the solution to a MWCO 30 KD Nanosep device, and remove reaction buffer and excess DNA via centrifugal ultrafiltration.

7. The AuNPs–MB conjugates (AuMB) should be washed 2 times via centrifugal ultrafiltration and resuspended in ultrapure water at a final concentration of 100 nM.

3.2 Preparation of QD–AuMBQ500 Composite Probe (See Note 3)

1. Add BHQ–2 amine solution to 200 μL 100 nM AuMB solution at a molar ratio of 500:1 to prepare AuMBQ500. Place the mixture in the dark for 1 h.
2. Purify the obtained AuMBQ500 by centrifugal ultrafiltration with a MWCO 30 KD Nanosep device for three times, and resuspend with 200 μL water.
3. Add 1 μL 1 μM SA–QDs solution and 5 μL Zonyl® FSN fluorosurfactant (1%) to 100 μL AuMBQ solution, and incubate in the dark for 30 min.

3.3 Labeling RSV gRNA in Living Cells with QD–AuMBQ500

1. Seed A549 cells in 35 mm glass bottom dishes, and place the dishes at 37 °C incubator in the presence of 5% CO_2.
2. Inoculate RSV to the cell monolayer at MOI 0.5 for 2 h at 37 °C when the cells reached 70–80% confluence, and maintain the cells for 48 h in the presence of 2% FBS at 37 °C.
3. Activate streptolysin O (SLO, 2×10^4 U/mL) with a newly prepared 20 mM cysteine solution for 10 min, and dilute SLO to 0.2 U/mL solution with DMEM.
4. Add QDAuMBQ solutions to SLO/DMEM solution to a final concentration of 1 nM.
5. Add the mixture to RSV infected A549 cells, and incubate it at 37 °C for 20 min.
6. Wash the cells with PBS for three times, and maintain the cells in DMEM for another 20 min to allow the cell membrane to recover.
7. Image the cells by fluorescence microscope to observe RSV gRNA.

3.4 Monitoring the RSV gRNA Replication Using QD–AuMBQ500 Probes in Live Cells

1. Seed A549 cells in 35 mm glass-bottom dishes, and place the dishes at 37 °C incubator in the presence of 5% CO_2.
2. Inoculate RSV to the cell monolayer at MOI 1 for 2 h at 37 °C when the cells reached 70–80% confluence, and maintain the cells in the presence of 2% FBS at 37 °C.
3. At each time point of 3, 6, 12, 24, 36, or 48 post-infection, label the viral gRNA in the infected cells with QD–AuMBQ500 in the presence of SLO as mentioned above, and then image the cells by fluorescence microscopy (*see* **Note 4**).

3.5 Imaging of Viral gRNA by iDFM

1. Seed A549 cells in 35 mm glass-bottom dishes, and place the dishes at 37 °C incubator in the presence of 5% CO_2.
2. Inoculate RSV to the cell monolayer at MOI 1 for 2 h at 37 °C when the cells reached 70–80% confluence, and maintain the cells for 48 h in the presence of 2% FBS at 37 °C.
3. Activate streptolysin O (SLO, 2×10^4 U/mL) with a newly prepared 20 mM cysteine solution for 10 min, and dilute SLO to 0.2 U/mL solution with DMEM.
4. Add QDAuMBQ solutions to SLO/DMEM solution to a final concentration with 1 nM.
5. Add the mixture to RSV infected A549 cells, and incubate at 37 °C for 20 min.
6. Wash the cells with PBS for three times, and maintain the cells in DMEM for another 20 min.
7. Fix cells with 4% paraformaldehyde for 30 min, and incubate the cells with 2% bovine serum albumin (BSA) at 37 °C for 1 h.
8. Mix solution A and solution B in the silver enhancer kit, and add the mixture to the dishes.
9. After 20 min, wash the dishes with distilled water, and fix by immersing in 2.5% sodium thiosulfate solution for 2 min.
10. Wash the dishes with distilled water again and image with iDFM.

4 Notes

1. The DNA sequences used to target RSV gRNA: Biotin–MB–SH (5′–Biotin– TATCTAACAAAGAAAAATGGGGCAAA TAACTTTGTATCTATSH –3′), blocking DNA (5′–HS–TAATCTAATT–3′).
2. SA–QDs solution and the prepared QD–AuMBQ500 should be stored in the dark at 4 °C.
3. Avoid diluting the SA–QDs solution before mixing with AuMBQ500, because the low concentration SA–QDs solution was unstable.
4. Image the living cells within 1 h after being labeled with QD–AuMBQ500; otherwise the nanoprobes might drop off from the gRNA as the replication of gRNA went on.

Acknowledgments

This work was supported by the National Natural Science Foundation of China (NSFC, No. 22134005 and 22322409) and

Chongqing Talents Program for Outstanding Scientists (cstc2021ycjh-bgzxm0178).

References

1. Su JR (2004) Emerging viral infections. Clin Lab Med 24(3):773–795
2. Kang J, Tahir A, Wang H et al (2021) Applications of nanotechnology in virus detection, tracking, and infection mechanisms. Wiley Interdiscip Rev Nanomed Nanobiotechnol 13(4):e1700
3. Cagno V, Andreozzi P, D'Alicarnasso M et al (2018) Broad-spectrum non-toxic antiviral nanoparticles with a virucidal inhibition mechanism. Nat Mater 17(2):195–203
4. Available from: https://www.who.int/emergencies/diseases/novel-coronavirus-2019
5. Liu S-L, Wang Z-G, Xie H-Y et al (2020) Single-virus tracking: from imaging methodologies to virological applications. Chem Rev 120(3):1936–1979
6. Pallas V, García JA (2011) How do plant viruses induce disease? Interactions and interference with host components. J Gen Virol 92: 2691–2705
7. Nikaeen G, Abbaszadeh S, Yousefinejad S (2020) Application of nanomaterials in treatment, anti-infection and detection of coronaviruses. Nanomedicine (Lond) 15(15): 1501–1512
8. Zheng LL, Li CM, Zhen SJ et al (2016) His-tag based in situ labelling of progeny viruses for real-time single virus tracking in living cells. Nanoscale 8(44):18635–18639
9. Brandenburg B, Lee LY, Lakadamyali M et al (2007) Imaging poliovirus entry in live cells. PLoS Biol 5(7):e183
10. Lakadamyali M, Rust MJ, Babcock HP et al (2003) Visualizing infection of individual influenza viruses. Proc Natl Acad Sci USA 100(16): 9280–9285
11. Lakadamyali M, Rust MJ, Zhuang X (2006) Ligands for clathrin-mediated endocytosis are differentially sorted into distinct populations of early endosomes. Cell 124(5):997–1009
12. Hollinshead M, Rodger G, Eijl HV et al (2001) Vaccinia virus utilizes microtubules for movement to the cell surface. J Cell Biol 154(2): 389–402
13. Rietdorf J, Ploubidou A, Reckmann I et al (2001) Kinesin-dependent movement on microtubules precedes actin-based motility of vaccinia virus. Nat Cell Biol 3(11):992–1000
14. Newsome TP, Scaplehorn N, Way M (2004) SRC mediates a switch from microtubule- to actin-based motility of vaccinia virus. Science 306(5693):124–129
15. Baltimore D (1971) Expression of animal virus genomes. Bacteriol Rev 35(3):235–241
16. Sivaraman D, Biswas P, Cella LN et al (2011) Detecting RNA viruses in living mammalian cells by fluorescence microscopy. Trends Biotechnol 29(7):307–313
17. Ewers H, Smith AE, Sbalzarini IF et al (2005) Single-particle tracking of murine polyoma virus-like particles on live cells and artificial membranes. Proc Natl Acad Sci USA 102(42):15110–15115
18. Wang S, Huang X, Huang Y et al (2014) Entry of a novel marine DNA virus, Singapore grouper iridovirus, into host cells occurs via clathrin-mediated endocytosis and macropinocytosis in a pH-dependent manner. J Virol 8(22):13047–13063
19. Liu S-L, Zhang Z-L, Tian Z-Q et al (2012) Effectively and efficiently dissecting the infection of influenza virus by quantum-dot-based single-particle tracking. ACS Nano 6(1): 141–150
20. Van Der Schaar HM, Rust MJ, Chen C, van der Ende-Metselaar H, Wilschut J, Zhuang X, Smit JM (2008) Dissecting the cell entry pathway of dengue virus by single-particle tracking in living cells. PLoS Pathog 4(12):e1000244
21. Boersma S, Rabouw HH, Bruurs LJM et al (2020) Translation and replication dynamics of single RNA viruses. Cell 183(7):1930–1945
22. Belza J, Opletalová A, Poláková K (2021) Carbon dots for virus detection and therapy. Mikrochim Acta 188(12):430
23. Zhang C-Y, Yeh H-C, Kuroki MT et al (2005) Single-quantum-dot-based DNA nanosensor. Nat Mater 4(11):826–831
24. Zheng LL, Yang XX, Liu Y et al (2014) In situ labelling chemistry of respiratory syncytial viruses by employing the biotinylated host-cell membrane protein for tracking the early stage of virus entry. Chem Commun 50(99): 15776–15779
25. Ceja-Fdez A, Carriles R, González-Yebra AL et al (2021) Imaging and SERS study of the au nanoparticles interaction with HPV and carcinogenic cervical tissues. Molecules 26(12): 3758

26. Zheng LL, Shuaia XJ, Liu Y et al (2022) One--donor-two-acceptors coupled energy transfer nanoprobe for recording of viral gene replication in living cells. Chem Eng J 434:134658
27. Li CM, Zheng LL, Yang XX et al (2016) DNA-AuNP networks on cell membranes as a protective barrier to inhibit viral attachment, entry and budding. Biomaterials 77:216–226
28. Karunatilaka KS, Filman DJ, Strauss M et al (2021) Real-time imaging of polioviral RNA translocation across a membrane. MBio 12(1):e03695-20
29. Santangelo PJ, Lifland AW, Curt P et al (2009) Single molecule-sensitive probes for imaging RNA in live cells. Nat Methods 6(5):347–349
30. Alonas E, Lifland AW, Gudheti M et al (2014) Combining single RNA sensitive probes with subdiffraction-limited and live-cell imaging enables the characterization of virus dynamics in cells. ACS Nano 8(1):302–315
31. Tyagi S, Kramer FR (1996) Molecular beacons: probes that fluoresce upon hybridization. Nat Biotechnol 14(3):303–308
32. Tyagi S, Bratu DP, Kramer FR (1998) Multicolor molecular beacons for allele discrimination. Nat Biotechnol 16(1):49–53
33. Wang K, Tang Z, Yang CJ et al (2009) Molecular engineering of DNA: molecular beacons. Angew Chem Int Ed Engl 48(5):856–870
34. Perlette J, Tan W (2001) Real-time monitoring of intracellular mRNA hybridization inside single living cells. Anal Chem 73(22):5544–5550
35. Saito Y, Mizuno E, Bag SS et al (2007) Design of a novel G-quenched molecular beacon: a simple and efficient strategy for DNA sequence analysis. Chem Commun 43:4492–4494
36. Yang CJ, Lin H, Tan W (2005) Molecular assembly of superquenchers in signaling molecular interactions. J Am Chem Soc 127(37): 12772–12773
37. Li N, Chang C, Pan W et al (2012) A multicolor nanoprobe for detection and imaging of tumor-related mRNAs in living cells. Angew Chem Int Ed Engl 51(30):7246–7230
38. Piao Y, Liu F, Seo TS (2012) A novel molecular beacon bearing a graphite nanoparticle as a nanoquencher for in situ mRNA detection in cancer cells. ACS Appl Mater Interfaces 4(12): 6785–6789
39. Kim Y-P, Oh Y-H, Oh E et al (2008) Energy transfer-based multiplexed assay of proteases by using gold nanoparticle and quantum dot conjugates on a surface. Anal Chem 80(12): 4634–4641
40. Draz MS, Fang BA, Li L et al (2012) Hybrid nanocluster plasmonic resonator for immunological detection of hepatitis B virus. ACS Nano 6(9):7634–7643
41. Gorshkov K, Susumu K, Chen J et al (2020) Quantum dot-conjugated SARS-CoV-2 spike pseudo-virions enable tracking of angiotensin converting enzyme 2 binding and endocytosis. ACS Nano 14:12234–12247

Chapter 13

A Near-Infrared Fluorescent Probe for Live-Cell Imaging of RNA in the Cancer Cells

Yong Liu and Zhe Chen

Abstract

Ribonucleic acid (RNA), a biomolecular compound polymerized by many riboside monomers, is one of the most essential substances in life. Monitoring RNA changes within organisms will play an important role in the diagnosis and treatment of major genetically related diseases. Herein, we synthesized a novel compound **QOT-NA** as near-infrared probe for imaging RNA in cancer cells. The synthesis method and photophysical properties of the probe were presented. The titration and selectivity experiments demonstrated that the probe detected RNA in vitro. Moreover, the RNA imaging experiments indicated that the near-infrared RNA probe **QOT-NA** can locate RNA within cancer cells.

Key words Near-infrared, Ribonucleic acid, Imaging, Cancer cells, Fluorescent probe

1 Introduction

Nucleic acid, a generic name for deoxyribonucleic acid (DNA) and RNA, is a biomolecular compound polymerized by many nucleotide monomers and one of the most essential substances in life [1]. Nucleic acid transcription plays an important role in gene expression [2]. For example, RNA can express genetic information on proteins in living biological systems [3]. Structural variations in RNA will lead to the occurrence of major genetically related diseases [4, 5]. Thus, monitoring RNA changes within organisms will play an important role in the diagnosis and treatment of major genetically related diseases. Tracking endogenous RNA in biological systems is necessary to understand the role of RNA in genetic expression [6].

The emergence of a series of low-molecular-weight RNA probes is sufficient to demonstrate the status of RNA in the biological research field. The RNA detection mechanism is mainly the self-assembly of molecular probes through hydrogen bonds with nucleic acid grooves because RNA is a spiral single-chain

Shin-ichi Sato (ed.), *Live-Cell RNA Imaging: Methods and Protocols*, Methods in Molecular Biology, vol. 2875,
https://doi.org/10.1007/978-1-0716-4248-1_13,

structure [7, 8]. However, the emission wavelengths of most RNA probes are located in the ultraviolet or visible range (300–650 nm) [9, 10], which limits their application in biological samples. Near-infrared probes with emission wavelengths greater than 650 nm have attracted much attention because it has light damages to biological samples and limited effect on the spontaneous background fluorescence signal of biomolecules [11, 12]. Thus, novel near-infrared RNA probes with large Stokes shift should be developed.

In this study, we modified quinoline with iodoethane to make it positively charged and then connected quinoline with triphenylethylene through furan and double bond to construct a large conjugated system. We synthesized a new compound **QOT-NA** as near-infrared RNA probe with large Stokes shift for live-cell imaging of RNA in cancer cells.

2 Materials

All the reagents and materials were obtained from commercial suppliers and were not purified further before the experiment. Unless otherwise specified, all solvents in this work should be analytically pure.

2.1 Reagents

1. 5-Formyl-2-furanboronic acid.
2. Iodoethane.
3. Bromotriphenylethylene.
4. Quinaldine.
5. $CuCl_2 \cdot 2H_2O$.
6. $NaAC \cdot 3H_2O$.
7. $MgCl_2$.
8. $NaNO_2$.
9. H_2O_2.
10. NaF.
11. $Na_3PO_4 \cdot 12H_2O$.
12. NaBr.
13. $NaClO \cdot 5H_2O$.
14. KNO_3.
15. KCl.
16. NaCl.
17. $NiCl_2 \cdot 6H_2O$.
18. $BaCl_2 \cdot 2H_2O$.
19. $SnCl_4$.

20. $ZnSO_4 \cdot 7H_2O$.
21. Na_2SO_4.
22. NaSCN.
23. $CaCl_2$.
24. Glutamic acid (Glu).
25. Cysteine (Cys).
26. Glucose (Glc).
27. Glutathione (GSH).
28. Dimethyl sulfoxide (DMSO).
29. N,N-dimethylformamide (DMF).
30. CH_3CN.
31. CH_3OH.
32. PBS buffer.
33. Tris–HCl buffer.
34. 10% formalin-fixed solution.
35. Dulbecco's Modified Eagle's Medium (DMEM).
36. Fetal bovine serum (FBS).
37. Penicillin–streptomycin liquid.
38. Triton ×-100.
39. Ribonuclease (RNase) powder.

2.2 Equipment

1. Labtech UV Power PC spectrometer.
2. HITACHI F4600 fluorescence spectrophotometer.
3. Olympus FV-1200 confocal laser scanning microscope.

2.3 RNA Detection in Vitro

2.3.1 Optical Properties of the Probe QOT-NA

1. Dissolve **QOT-NA** in DMF to obtain 1 mM stock solution.
2. Use **QOT-NA** (5 μM) in photophysical experiments. Detect absorption spectra and fluorescence responses of **QOT-NA** in DMSO, DMF, H_2O, CH_3CN, CH_3OH, PBS, and Tris–HCl buffer solution. The absorption wavelength of **QOT-NA** should be used as maximum absorption wavelength in different solvents (*see* **Note 1**).

2.3.2 Preparation of RNA and DNA Solutions

1. Dissolve 5 mg of RNA or DNA in Tris–HCl buffer to obtain 0.1 mg/mL stock solution (*see* **Notes 2** and **3**).
2. The solution concentration should be determined by Lambert–Beer law (*see* **Note 4**).
3. Dilute RNA/DNA to various concentrations in RNA/DNA titration experiment (*see* **Note 5**).

4. Prepare the solutions of various biological analytes from $CuCl_2·2H_2O$, NaAC·$3H_2O$, $MgCl_2$, $NaNO_2$, H_2O_2, NaF, $Na_3PO_4·12H_2O$, NaBr, NaClO·$5H_2O$, KNO_3, KCl, NaCl, $NiCl_2·6H_2O$, $BaCl_2·2H_2O$, $SnCl_4$, $ZnSO_4·7H_2O$, Na_2SO_4, NaSCN, $CaCl_2$, glutamic acid (Glu), cysteine (Cys), glucose (Glc), and glutathione (GSH). [GSH, Cys, Glu] = 0.1 mM, [other] =1 mM.
5. Dilute the test solutions to 1 mM with Tris–HCl buffer solution (GSH, Cys, and Glu were diluted to 0.1 mM).
6. Take 1.99 mL of the diluents into the test tube, and add with 10 μL of probe **QOT-NA** (1 mM). After mixing evenly, a series of probe test solutions with different test objects should be obtained.

2.4 RNA Imaging In Vivo

1. Culture HeLa cells in a sterile constant-temperature incubator at 37 °C and 5% CO_2.
2. Fix the cells by treating live HeLa cells with 4% formalin-fixed solution for 30 min.
3. Cell culture medium: Dulbecco's Modified Eagle's Medium (DMEM), fetal bovine serum (FBS), and penicillin–streptomycin liquid in the ratio of 89:10:1 (store at 4 °C).
4. 0.5% Triton ×-100 diluted with PBS (store at 4 °C).
5. Dissolve 30 mg of RNase powder in 1 mL of PBS (about 30 mg/mL RNase stock solution, store at −4 °C).

3 Methods

3.1 Near-Infrared Probe Synthesis

3.1.1 Synthesis of Probe QOT-NA

1. Dissolve compounds **1** (350 mg, 1.0 mmol) and **2** (299 mg, 1.0 mmol) in 30 mL of methanol.
2. The mixture should be added with 0.2 mL of piperidine and stirred at 25 °C for 2 h under the protection of nitrogen.
3. Remove methanol by vacuum distillation after the reaction.
4. Purify the resulting solid by a silica gel column (DCM, MeOH = 100:1). About 172 mg of **QOT-NA** brown powder was obtained (Scheme 1, yield 27%). ^{1}H NMR (400 MHz, DMSO-d6) δ 8.95 (d, J = 8.9 Hz, 1H), 8.53–8.39 (m, 2H), 8.31 (d, J = 7.9 Hz, 1H), 8.17 (t, J = 7.5 Hz, 1H), 8.05 (d, J = 15.3 Hz, 1H), 7.92 (t, J = 7.4 Hz, 1H), 7.40 (s, 3H), 7.25 (d, J = 13.9 Hz, 7H), 7.14 (s, 4H), 6.98 (d, J = 5.6 Hz, 2H), 6.32 (d, J = 15.3 Hz, 1H), 6.12 (d, J = 3.5 Hz, 1H), 4.75 (s, 2H), 1.43 (t, J = 6.8 Hz, 3H). ^{13}C NMR (101 MHz, DMSO-d6) δ 159.99, 154.48, 150.60, 144.67, 144.04, 142.60, 139.97, 138.54, 135.57, 133.25, 131.74, 131.11, 130.79,

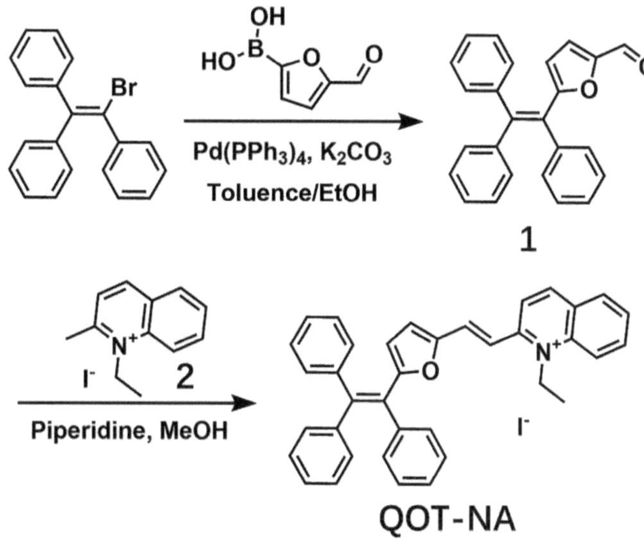

Scheme 1 Synthesis of probe **QOT-NA**

130.14, 129.29, 128.64, 128.31, 128.15, 121.93, 121.02, 119.13, 116.98, 113.47, 46.77, 14.19. HRMS (m/z): $[M\text{-}I]^+$ calcd for $C_{37}H_{30}NO^+$, 504.2322; found, 504.2325.

3.2 Spectral Measurements

1. In the UV and fluorescence spectra test, the probe stock solution should be diluted to 5 μM to test the optical properties of the probe in diverse solutions (*see* **Note 6**).
2. Fluorescence quantum yield (*see* **Note** 7).

   The fluorescence quantum yields should be calculated by the following equation:

$$\phi_s = \phi_r \left(\frac{A_r(\lambda_r)}{A_s(\lambda_s)} \right) \left(\frac{n_s^2}{n_r^2} \right) \frac{F_s}{F_r}$$

   In the above equation, **s** and **r** stand for sample and reference, respectively, and ϕ and **F** stand for quantum yield and integrated emission intensity, respectively. **A** and **n** stand for absorbance and refractive index, respectively.
3. Dissolve RNA and DNA in Tris–HCl buffer solution (*see* **Note 8**). Determine the solution concentration by Lambert–Beer law.
4. Add 10 μL of the probe stock solution to a series of test tubes for the titration experiments of nucleic acids. The tubes should be added with different volumes of nucleic acid stock solution to obtain series of probes/nucleic acid mixed solutions from high to low proportions (*see* **Note 9**, Fig. 1). RNA and DNA titration fluorescence spectra should be obtained under 540 nm excitation.

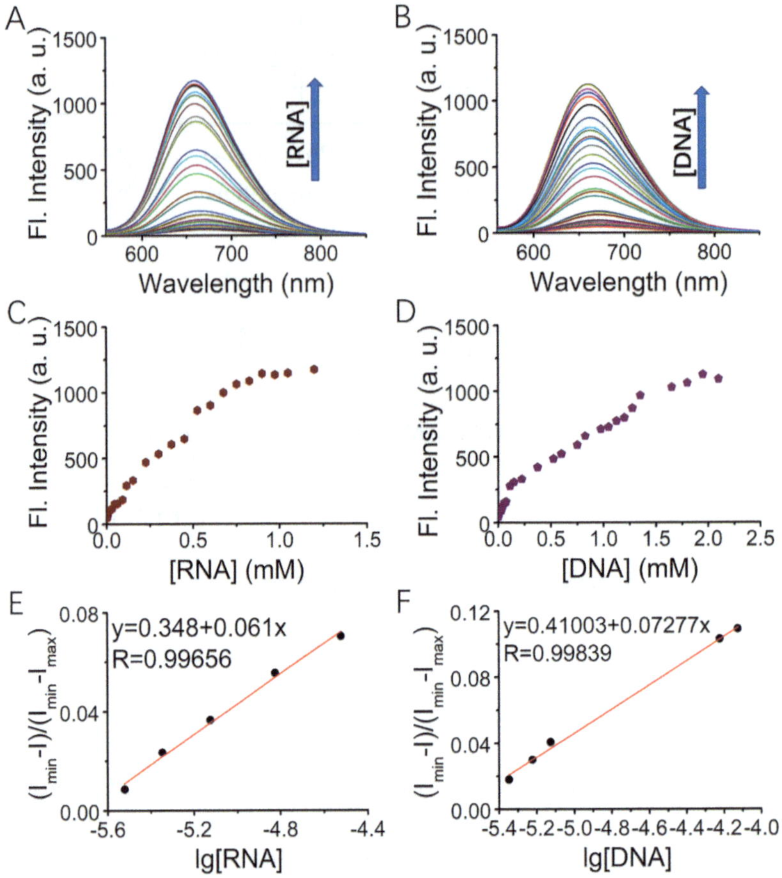

Fig. 1 Fluorescence spectra of **QOT-NA** upon addition of RNA (**a**) and DNA (**b**) in Tris–HCl buffer solution. Fluorescence variation of the probe **QOT-NA** to RNA (**c**) and DNA (**d**) at 660 nm. The limit of detection of **QOT-NA** for RNA (**e**) and DNA (**f**). $\lambda_{ex} = 540$ nm

5. Prepare 10 mM concentration of various analyte stock solution for selectivity experiment (Fig. 2). The solution should be diluted to 1 mM for selectivity experiment. The test solutions should be diluted to 1 mM with Tris–HCl buffer solution.
6. Prepare **QOT-NA** solution for selectivity experiment by mixing 1.99 mL of the diluent and 10 μL probe **QOT-NA** (1 mM). After mixing evenly, a series of probe test solutions with different test objects should be obtained.

3.3 Cell Culture and Imaging

1. Before imaging, the cells should be incubated with the culture medium containing **QOT-NA** (5 μM) for 30 min and then washed twice with PBS (*see* **Note 10**). The clean culture medium should be added and imaged with microscope.
2. For RNase digestion imaging experiment (*see* **Note 11**, Fig. 3), the fixed cells should be pierced with 0.5% Triton ×-100 for

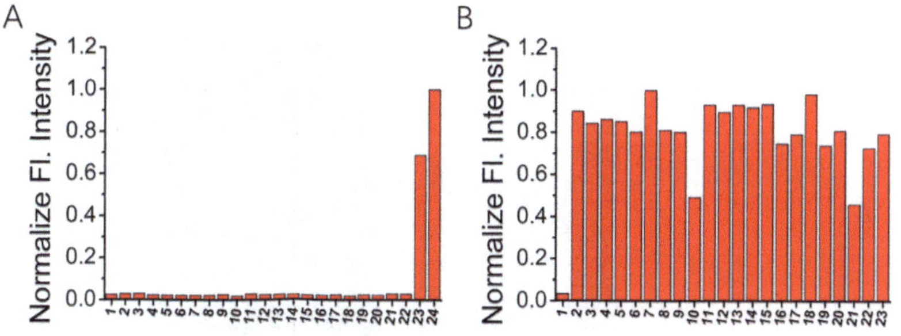

Fig. 2 (**a**) Fluorescence intensity at 660 nm of **QOT-NA** in different analytes. From left to right: (1) Blank, (2) AC^-, (3) Br^-, (4) Ca^{2+}, (5) Cl^-, (6) Cys, (7) F^-, (8) Glu, (9) GSH, (10) H_2O_2, (11) K^+, (12) Mg^{2+}, (13) Na^+, (14) NO_2^-, (15) NO_3^-, (16) ClO^-, (17) PO_4^{3-}, (18) Glc, (19) SCN^-, (20) Sn^{4+}, (21) SO_4^{2-}, (22) Zn^{2+}, (23) DNA, (24) RNA. (**b**) Fluorescence intensity at 660 nm of **QOT-NA** upon addition of RNA and various analytes. From left to right: (1) Blank, (2) RNA, (3) AC^-&RNA, (4) Br^-&RNA, (5) Ca^{2+}&RNA, (6) Cl^-&RNA, (7) Cys&RNA, (8) F^-&RNA, (9) Glu&RNA, (10) GSH&RNA, (11) H_2O_2&RNA, (12) K^+&RNA, (13) Mg^{2+}&RNA, (14) Na^+&RNA, (15) NO_2^-&RNA, (16) NO_3^-&RNA, (17) ClO^-&RNA, (18) PO_4^{3-}&RNA, (19) Glc&RNA, (20) SCN^-&RNA, (21) Sn^{4+}&RNA, (22) SO_4^{2-}&RNA, (23) Zn^{2+}&RNA. **[QOT-NA]** = 5 μM. [GSH, Cys, Glu] = 0.1 mM, [other] =1 mM. λ_{ex} = 540 nm

2 min and washed with PBS (*see* **Note 12**). The cells should be then divided into two groups: λ_{ex} = 543 nm and λ_{em} = 600–700 nm.

3. Treat the cells in the experimental group with PBS containing RNase (30 mg/mL) for 2 h. After washing again, the two groups of fixed HeLa cells should be incubated with the probe **QOT-NA** and imaged. λ_{ex} = 543 nm. λ_{em} = 600–700 nm.
4. Evaluate in situ imaging characteristics of RNA in living cells (*see* **Note 13**). In the receiving range of 400–790 nm, the range of fluorescence signal is 590–710 nm, which is basically consistent with the experimental results of RNA titration in spectral experiment. λ_{ex} = 543 nm. λ_{em} = 400–790 nm.
5. Photostability: Incubate the cells with **QOT-NA** for 30 min, wash twice with PBS, and then scan continuously with a microscope for 10 min. λ_{ex} = 543 nm. λ_{em} = 600–700 nm.

4 Notes

1. This part mainly studies the photophysical properties of the probe to verify whether it is a near-infrared fluorescent compound. All fluorescent spectra should be obtained using the maximum absorption wavelength as the excitation wavelength in various solvents.

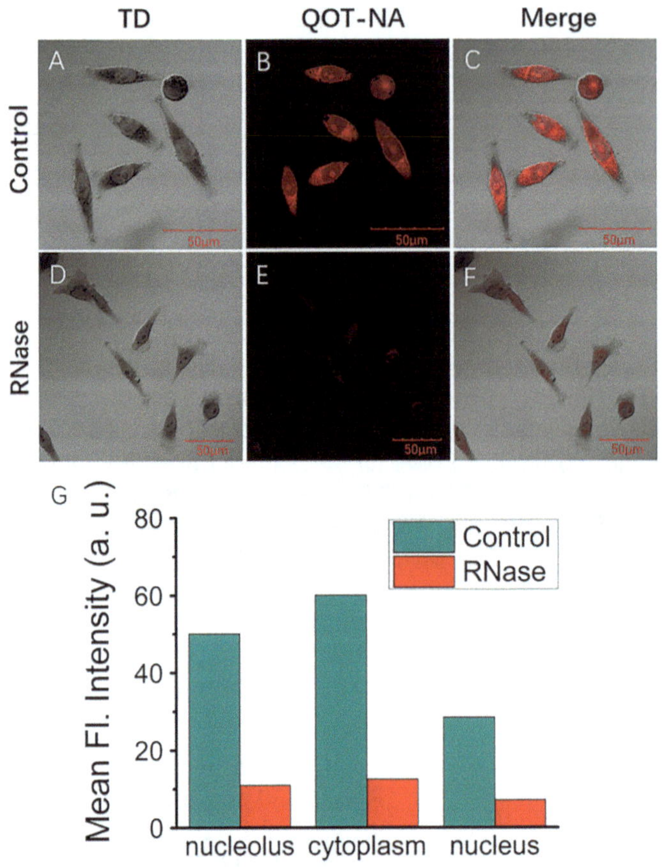

Fig. 3 (**a**–**c**) Images of HeLa cells treated with **QOT-NA**, (**d**–**f**) images of HeLa cells treated with **QOT-NA** and RNase. (**g**) Mean fluorescence intensity of the nucleolus, cytoplasm, and nucleus before and after treatment with RNase. $\lambda_{ex} = 543$ nm. $\lambda_{em} = 600$–700 nm. Bar = 50 μm

2. This concentration could not be used as a test concentration with the probe. It should be converted to a volume-molar concentration by the Lambert–Beer law.

3. RNA and DNA are easily hydrolyzed in the aqueous phase and thus should be tested timely after matching the solution.

4. RNA/DNA solution: RNA/DNA concentration should be calculated by the following equation. The ultraviolet absorption intensity is stranded due to the size of the electron energy-level transition probability and abides by the Lambert–Beer law (1).

$$A = -\log \frac{I}{I_0} = \xi cl \tag{1}$$

A stands for absorbance, ε stands for extinction coefficient, extinction coefficient of RNA is 7700 (DNA is 6600), *c* stands for molar concentration, *l* stands for length of sample pool, and I_0 and I stand for intensity of the incident light and transmission light, respectively.

5. For prevention of the hydrolysis of nucleic acids, such solutions should be tested in time.
6. The excitation wavelength is 540 nm, and the excitation and emission slit widths are 5 and 5 nm, respectively. Unless otherwise noted, for the titration and selectivity experiments, the excitation wavelength is 540 nm, and the excitation and emission slit widths of fluorescence spectrophotometer are 5 and 5 nm, respectively.
7. The quantum yield of the probe should be calculated to determine whether the probe has the capability of fluorescence imaging.
8. To simulate the intracellular environment, all test solutions used Tris–HCl buffer solutions [13, 14].
9. About 1.99 mL of the diluent is taken into the test tube and added with 10 μL **QOT-NA** (1 mM) to each test tube. After mixing evenly, a series of probe test solutions in different concentrations of RNA/DNA is obtained.
10. Cleaning off the excess probes to avoid the precipitated probes to attach to the cell surface and confers the imaging effect.
11. For RNA digestive experiments: First, the two identical sets of HeLa cells are cultivated. The two-disc cells are fixed and perforated under the same conditions. Second, one plate is processed with RNase for 2 hours and another as a control experiment. Third, the two-disc cells are restrained with 5 μM **QOT-NA** for 30 min and finally subjected to cell imaging experiments.
12. All fixed cells are pierced with 0.5% Triton ×-100 for 2 min in RNA digestive experiments. The cell treatment time with 0.5% Triton ×-100 should not be too long.
13. In situ imaging: Living cells are incubated with 5 μM **QOT-NA** for 30 min and subjected to imaging experiment. The emission bands used are different from the other imaging experiments of **QOT-NA**.

References

1. Liu Y, Meng F, He L, Yu X, Lin W (2016) Fluorescence behavior of a unique two-photon fluorescent probe in aggregate and solution states and highly sensitive detection of RNA in water solution and living systems. Chem Commun 52:8838–8841
2. Niu J, Liu Y, Wang W, Lin W (2019) A novel mitochondria-targetable probe for imaging endogenous deoxyribonucleic acid in biological systems. J Photoch Photobio A 378:57–65
3. Mattick JS (2004) RNA regulation: a new genetics? Nat Rev Genet 5:316–323
4. Kataoka K, Miyoshi H, Sakata S, Dobashi A, Couronne L, Kogure Y, Sato Y, Nishida K, Gion Y, Shiraishi Y (2019) Frequent structural

variations involving programmed death ligands in Epstein-Barr virus-associated lymphomas. Leukemia 33:1687–1699
5. Hardy J, Singleton A (2009) Current concepts genomewide association studies and human disease. N Engl J Med 360:1759–1768
6. Liu Y, Niu J, Wang W, Lin W (2019) Tracking of mitochondrial endogenous ribonucleic acid in the cancer cells and macrophages using a novel small-molecular fluorescent probe. Anal Chem 91:1715–1718
7. Bhatia D, Sharma S, Krishnan Y (2011) Synthetic, biofunctional nucleic acid-based molecular devices. Curr Opin Biotechnol 22:475–484
8. Krishnan Y, Simmel FC (2011) Nucleic acid based molecular devices. Angew Chem Int Edit 50:3124–3156
9. Liu Y, Niu J, Wang W, Ma Y, Lin W (2018) Simultaneous imaging of ribonucleic acid and hydrogen sulfide in living systems with distinct fluorescence signals using a single fluorescent probe. Adv Sci 5:1700966
10. Wang Z, Wang C, Fang T, Liu Y (2022) Constructing a novel near-infrared fluorescent probe with a large stokes shift for detection of RNA in the cancer cells. Dyes Pigments 200: 110126
11. Wang Z, Liu Y, Wang W, Zhao C, Lin W (2021) A single small molecule fluorescent probe for imaging RNA distribution and detecting endogenous SO_2 through distinct fluorescence channels. New J Chem 45: 19812–19817
12. Yuan L, Lin W, Zheng K, He L, Huang W (2013) Far-red to near infrared analyte-responsive fluorescent probes based on organic fluorophore platforms for fluorescence imaging. Chem Soc Rev 42:622–661
13. Yang Z, Zhao N, Sun Y, Miao F, Liu Y, Liu X, Zhang Y, Ai W, Song G, Shen X (2012) Highly selective red- and green-emitting two-photon fluorescent probes for cysteine detection and their bio-imaging in living cells. Chem Commun 48:3442–3444
14. Sun YQ, Wang P, Liu J, Zhang J, Guo W (2012) A fluorescent turn-on probe for bisulfite based on hydrogen bond-inhibited CN isomerization mechanism. Analyst 137:3430–3433

Chapter 14

A Fluorogenic Sensor via Catalytic Hairpin Assembly for Precise Live-Cell Imaging of mRNA

Su-Jing Zhao, Zhenkun Wu, and Jian-Hui Jiang

Abstract

Accurate evaluation of the tumor-related mRNA expression levels provides important information for cancer diagnosis and therapy. We recently reported a fluorogenic sensor by coupling tetrazine-mediated bioorthogonal reaction with catalytic hairpin assembly for precise imaging of GalNac-T mRNA in live cells. Fluorescence signals are specifically generated by target mRNA triggered spatial localization of bioorthogonal chemicals. The proposed fluorogenic sensor exhibits low detection limit with signal amplification process. Importantly, the method can avoid false-positive results in biological environment because of the "click to release" feature of bioorthogonal reaction. Our method may provide a useful tool for precise imaging of intracellular mRNA with low expression levels.

Key words Bioorthogonal reaction, Catalytic hairpin assembly, mRNA imaging, Biosensor

1 Introduction

The ability of low-abundance biomarker detection is essential for clinical diagnostic of various diseases including cancer. To date, several DNA circuits, such as hybridization chain reaction (HCR) [1–5], entropy-driven catalysis (EDC) [6–10], and catalytic hairpin assembly (CHA) [11–15], have been utilized to construct amplified biosensors for sensitive RNA detection and imaging. Despite the important advances, these biosensors are commonly functionalized with a fluorophore and quencher to generate fluorescence signals, which may be degraded by endogenous nucleases in a cellular environment to lead to high background signals [16, 17]. Methods allowing accurate and sensitive imaging of intracellular mRNAs are still highly demanded. We recently reported a fluorogenic sensor for precise and amplified imaging of mRNA in live cells [18]. Tetrazine-mediated bioorthogonal reactions are coupled with CHA circuits to construct the fluorogenic sensors. Target mRNA triggered CHA results in a spatial proximity of bioorthogonal chemicals to

Shin-ichi Sato (ed.), *Live-Cell RNA Imaging: Methods and Protocols*, Methods in Molecular Biology, vol. 2875,
https://doi.org/10.1007/978-1-0716-4248-1_14,

yield high effective concentrations, allowing efficient decaging of vinyl ether group to regenerate the fluorescence signals. We here describe the synthesis of two bioorthogonal chemicals, tetrazine (Tz) and vinyl ether caged pro-fluorophore (PF), DNA conjugation, and fluorescence detection of mRNA in vitro and in live cells. Our method may provide a powerful platform for bioorthogonal chemistry studies and biomolecule detection.

2 Materials

All chemicals are analytical grade and used directly without further purification. All solutions are prepared using ultrapure water obtained from a Milli-Q system (resistance>18.2 MΩ cm^{-1}).

2.1 Chemical Reagents

1. 4-Szidobutyricacid.
2. Thionyl chloride.
3. Fluorescein.
4. 1-(3-Hydroxyphenyl) piperazine.
5. Vinylboronicanhydride pyridine complex.
6. Nickel (II) trifluoromethanesulfonate.
7. Hydrazine monohydrate.
8. 3-Cyanobenzyl alcohol.
9. Phosphorus tribromide.
10. Sodium azide.
11. Magnesium sulfate.
12. Acetic acid.
13. Triethylamine.
14. Dimethyl sulfoxide (DMSO).
15. Trifluoroacetic acid (TFA).
16. N,N-diisopropylethylenediamine (DIPEA).
17. Potassium hydroxide (KOH).
18. Copper (II) acetate.
19. Pyridine.
20. Tetrahydrofuran (THF).
21. Dichloromethane (DCM).
22. Acetic acid ethyl ester (EtOAc).
23. Methanol (MeOH).
24. Acetonitrile (MeCN).
25. Hydrochloric acid (HCl).
26. Deuterium dimethyl sulfoxide (DMSO-d_6).

27. Chloroform-d ($CDCl_3$).
28. Hoechst 33342.
29. 30% Acrylamide/N,N-methyldiacrylamide solution (Acr/Bis).
30. Ammonium persulfate (APS).
31. N,N,N,N-Tetramethylenediamine (TEMED).
32. 10,000× 4S GelRed solution.
33. 10× DNA loading Buffer.

2.2 Oligonucleotides (5′-3′)

1. H1: GTGTCTTATGCGGATAGTGAAAGCCCATGTGCCG CGGCTTTCACTA/iDBCOdT/TCCGCA.
2. H2:(DBCO) AGTGAAAGCCGCGGCACATGGGCTTT CACTATCCGCATACCATGTGCCGCGG.
3. GalNac-T mRNA mimics: GCTTTCACTATCCGCATAAGA CAC.
4. TK1 mRNA mimics: TGATCAAGTATGCCAAAGACAC.
5. MnSOD mRNA mimics: AATCAACTGGGAGAATGTAACT.
6. C-myc mRNA mimics: CCTCAACGTTAGCTTCACCAA.
7. Survivin mRNA mimics: TCTCAAGGACCACCGCATCTCT.

2.3 Buffers

1. 0.1 M triethylamine–acetic acid buffer (TEAA): 0.1 M triethylamine, 0.1 M acetic acid, pH 7.0.
2. 1× PBS: 10 mM phosphate, 137 mM NaCl, 2.7 mM KCl, pH 7.4.
3. 1× TBE: 89 mM Tris, 89 mM boric acid, 2 mM EDTA, pH 8.0–8.6.

2.4 Biological Reagents

1. Dulbecco's modified Eagle medium.
2. RPMI 1640 medium.
3. Penicillin–streptomycin.
4. Fetal bovine serum.
5. Lipofectamine 3000.
6. Opti-MEM.
7. Trypsin–EDTA (0.25%).

2.5 Instruments

1. High-performance liquid chromatography.
2. C18 column (5 μm, 150 × 3.0 mm).
3. Fluorescence spectrometer.
4. Gel imaging system.
5. Microplate reader.
6. Mass spectrometry.

7. Confocal laser scanning microscope.
8. Freeze dryer.
9. Nanodrop One spectrophotometer.
10. Electrophoresis system.
11. CO_2 incubator.

3 Methods

3.1 Synthesis of Vinyl Ether Caged Rhodol Pre-fluorophore (PF)

3.1.1 The General Synthetic Route Is Illustrated in Scheme 1

Compound 1 Fresh distilled thionyl chloride (4.37 mL, 60 mmol) is slowly added to 4-azidobutyric acid (50 mmol) at 0 °C via syringe. The solution is kept at room temperature for 15 h and then kept at 40 °C for another 1 h. The excess of $SOCl_2$, HCl, and SO_2 are evaporated. The crude product is directly used without further purification.

Compound 2 Fluorescein (6.65 g, 20 mmol) and KOH (22.44 g, 400 mmol) in water (30 mL) are heated to reflux for 5 h. The reaction mixture is then cooled to 0 °C, acidified with HCl, and extracted with ethyl ether. The extract is next concentrated, crystallized in DCM to give compound 2 (4.38 g, 85%).

Compound 3 Compound 2 (1.24 g, 4.8 mmol) and 1-(3-hydroxyphenyl)-piperazine (853 mg, 4.8 mmol) are added to the pressure tube and dissolved in TFA (20 mL). The reaction is stirred at

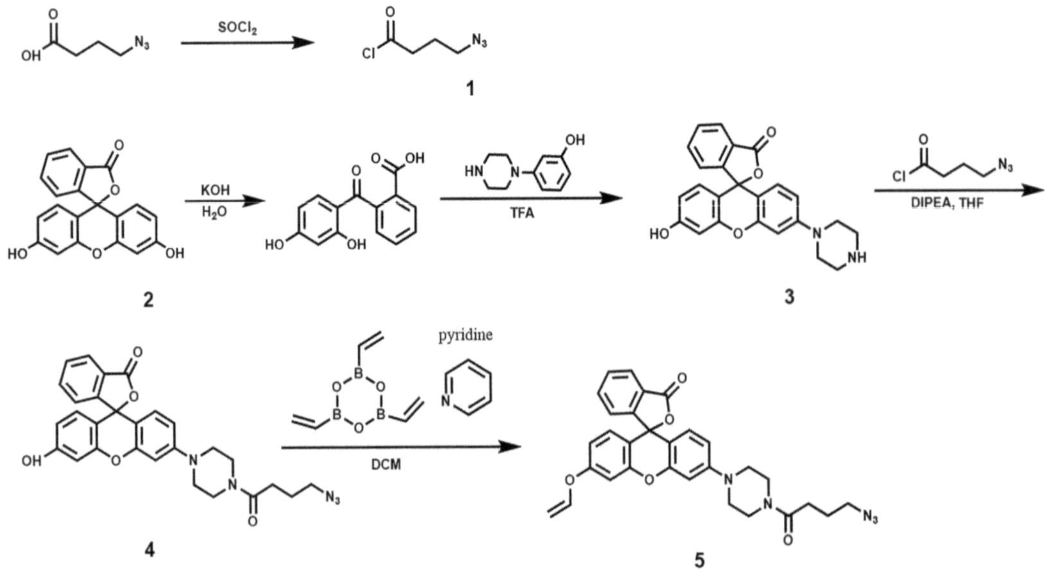

Scheme 1 Synthetic route of vinyl ether caged rhodol pre-fluorophore. (Reproduced from Ref. 18 with permission from ACS)

95 °C for 3 h. After cooling, the reaction mixture is poured into ethyl ether. The resulting precipitate is collected, immediately redissolved in MeOH, and then evaporated under reduced pressure to give a red solid. The crude product is purified by silica column chromatography using MeOH-CH_2Cl_2 1:1 (v/v) as the eluent to yield compound 3 as a red solid (1.35 g, 70%).

Compound 4 Compound 1 (191.8 mg, 1.3 mmol) in DCM and N, N-diisopropyl ethylamine (116.9 mg, 1.65 mmol) are dropwise added into compound 3 (508.7 mg, 1.5 mmol) in THF. The solution is stirred at room temperature for 2 h and then evaporated under reduced pressure to give a red solid. The crude product is purified by silica column chromatography using MeOH-CH_2Cl_2 1:10 (v/v) as the eluent to yield compound 4 as a red solid (655 mg, 86.7%).

Compound 5 Copper (II) acetate (270.6 mg, 1.49 mmol) is stirred at room temperature in anhydrous DCM (15 mL) for 10 min. Compound 4 (762.8 mg, 1.49 mmol), vinylboronicanhydride pyridine complex (223.5 mg, 1.49 mmol), and pyridine (0.46 mL, 14.9 mmol) are then added, and the reaction is stirred at room temperature for 48 h. The reaction mixture is extracted with EtOAC. The organic layers are washed with 2 M HCl, dried over magnesium sulfate, and concentrated in vacuo. The crude product is purified by silica column chromatography using MeOH-CH_2Cl_2 1:10 (v/v) as the eluent to give compound 5 as a yellow solid (524 mg, 65.5%).

3.2 Synthesis of Tetrazine (Tz)

3.2.1 The General Synthetic Route Is Illustrated in Scheme 2

Compound 6 3-cyanobenzyl alcohol (460 mg, 3.76 mmol), nickel (II) trifluoromethanesulfonate (268 mg, 0.75 mmol), hydrazine monohydrate (11.4 mL, 188 mmol), and acetonitrile (1.96 mL, 37.6 mmol) are added to a flask under nitrogen. The mixture is stirred at 60 °C for 18 h. Then the reaction mixture is cooled to room temperature, and 5 mL sodium nitrite solution (1.40 g, 18.8 mmol) is slowly added. Afterward, 1 M HCl solution is slowly added to reach pH = 3. The mixture is extracted with DCM, and the organic phase is dried over magnesium sulfate. The crude product is purified using silica column chromatography using CH_2Cl_2-EtOAC 2:1 (v/v) as the eluent to give compound 6 as a purplish red solid (303.9 mg, 40%).

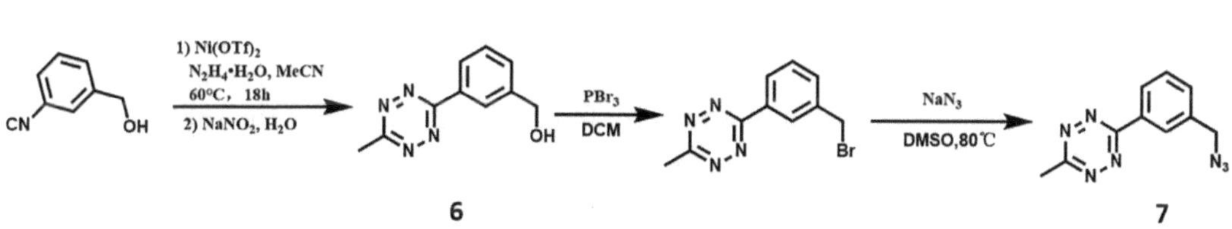

Scheme 2 Synthetic route of tetrazine. (Reproduced from Ref. 18 with permission from ACS)

Compound 7 Phosphorous tribromide (0.24 mL, 2.5 mmol) is dropwise added to a stirred suspension of compound 6 (1.03 g, 5.1 mmol) under nitrogen at 0 °C in DCM (25 mL). After 15 min, the mixture is warmed to room temperature and stirred for another 2 h, and then the reaction mixture is quenched with H_2O (15 mL). The aqueous phase is extracted with DCM (30 mL × 3), and the organic layer is then washed with brine (20 mL × 3), dried over magnesium sulfate, and concentrated in vacuo. The crude product (525.1 mg) and sodium azide (387 mg, 5.97 mmol) are heated to 80 °C in DMSO (10 mL) overnight. After 12 h, the reaction is quenched with water (30 mL × 3), and the product is extracted with EtOAC. The organic layer is washed with brine (20 mL × 3), dried over magnesium sulfate, and concentrated in vacuo to yield compound 7 as a purplish red solid (368.62 mg, 31.8%).

3.3 Oligonucleotide Modification

1. Dibenzocyclooctyne (DBCO)-modified DNA hairpins (H1 and H2) are dissolved in 200 μL double-distilled H_2O (ddH_2O) to reach a final concentration of 20 μM, respectively.
2. Equal volumes of Tz and PF in DMSO with a final concentration of 20 mM are added into the above solution of H1 and H2, respectively.
3. Solutions are stirred at room temperature for 6 h (*see* **Note 1**) to ensure efficient click conjugation to generate H1-Tz and H2-PF, respectively.
4. The prepared DNA products are then lyophilized in a vacuum freeze dryer to remove the solvent (*see* **Note 2**).
5. The freeze-dried products of H1-Tz and H2-PF are then redissolved in 50 μL ddH_2O for further HPLC purification.

3.4 Oligonucleotide Purification

The crude products of H1-Tz and H2-PF are purified using an Agilent 1260 HPLC system with a C18 Column (150 × 3.0 mm, 5 μm).

1. 0.1 M TEAA buffer is used as mobile phase A, and MeCN is used as mobile phase B. Separation gradient is set as follows: 0–5 min, 95% 0.1 M TEAA, 5% CH_3CN; 5–25 min, 95–55% 0.1 M TEAA, 5–45% CH_3CN; 25–80 min, 55–5% 0.1 M TEAA, 45–95% CH_3CN. The flow rate is set as 1 mL min^{-1}; and detection wavelength is 260 nm.
2. Products with retention time of 18.3 min are collected for H1-Tz, and products with retention time of 29.4 min are collected for H2-PF. The collected solutions are then lyophilized in a vacuum freeze dryer to remove the solvent.
3. The freeze-dried products of H1-Tz and H2-PF are redissolved in 30 μL ddH_2O, respectively (*see* **Note 3**). The concentrations

of modified DNA hairpins are determined using a Nanodrop one UV-Vis spectrophotometer. Modified DNA hairpins are then kept at 4 °C for further use.

3.5 In Vitro Detection of mRNA

The working principle of the fluorogenic sensor for mRNA detection is schematically illustrated in Scheme 3. In the presence of target mRNA, CHA is triggered to generate DNA duplexes, resulting in spatial proximity of two bioorthogonal chemicals (Scheme 3a). Such a spatial localization could greatly increase the local concentration of chemicals, allowing efficient removal of vinyl ether group to regenerate the fluorescence signals (Scheme 3b).

All the DNA reactants are treated as follows to ensure the correct folding before use: H1-Tz and H2-PF (2 μM) in 1× PBS buffer are heated to 95 °C for 5 min, then left on ice for 5 min, and finally left at room temperature for another 2 h.

3.5.1 Native Polyacrylamide Gel Electrophoresis

1. Add 4 mL of 30% Acr-Bis solution to 6 mL ddH_2O, and mix to obtain a 12% PAGE gel solution. Add 8 μL fresh 10% APS solution and 10 μL TEMED into 10 mL 12% PAGE gel solution, and mix immediately at room temperature (*see* **Note 4**).

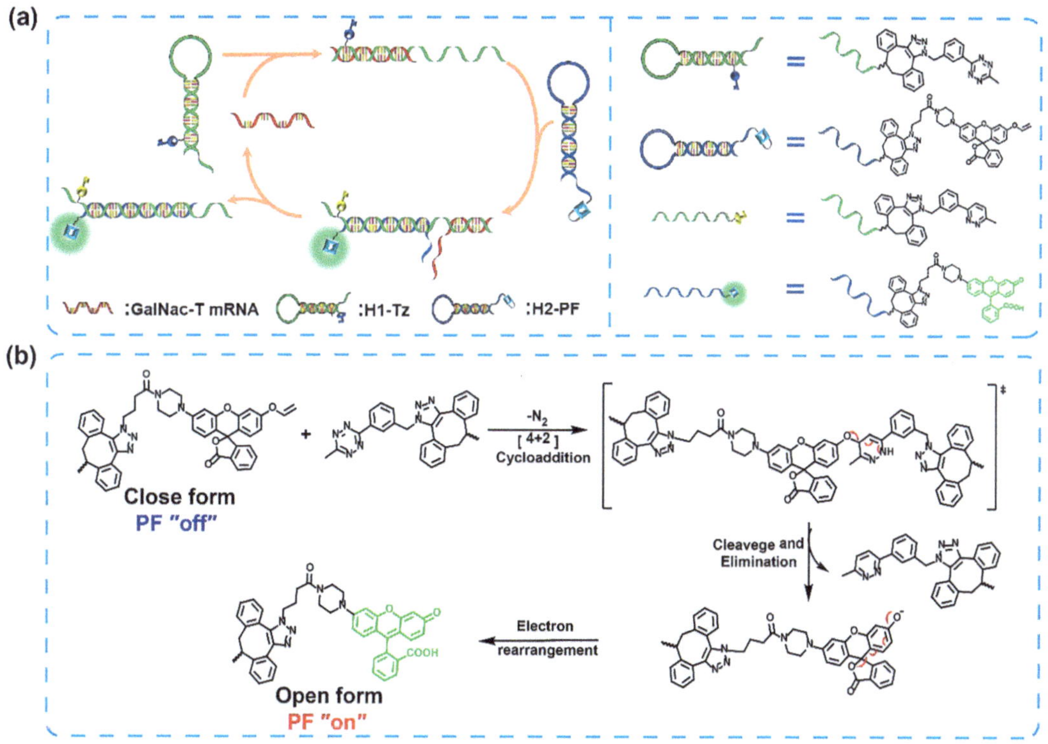

Scheme 3 Illustration of proximity-driven bioorthogonal reactions via catalytic hairpin assembly for mRNA imaging. (**a**) Working principle of the fluorogenic sensor. (**b**) Detailed structures of H1-Tz and H2-PF. (Reproduced from Ref. 18 with permission from ACS)

2. Sample preparation: (a) 1 μM H1; (b) 1 μM H2; (c) 1 μM H1 + 1 μM H2; (d) 1 μM H1 + 1 μM H2 + 10 nM GalNac-T mRNA mimics. Reaction volume is 10 μL. Samples are incubated at 37 °C for 4 h and then added 1 μL 10 × DNA loading buffer, respectively.
3. Add 10 μL to the above prepared samples in the selected lane of the gel, respectively. The electrophoretic voltage is 90 mV, and the running time is 1 h.
4. After electrophoresis, the gel is stained in 1× 4S GelRed solution (diluted with 1× TBE buffer) for 10 min and imaged by a gel imaging system.

3.5.2 *In Vitro Fluorescence Response*

Feasibility Assay

1. Four samples are prepared as follows: (a) 100 nM H1-Tz in 100 μL 1× PBS buffer; (b) 100 nM H2-PF in 100 μL 1× PBS buffer; (c) 100 nM H1-Tz and H2-PF in 100 μL 1× PBS buffer; (d) 10 nM target mRNA mimics, 100 nM H1-Tz and H2-PF in 100 μL 1× PBS buffer. All samples are incubated at 37 °C for 4 h.
2. Fluorescence spectra are then collected at room temperature. The excitation wavelength is 510 nm, and the fluorescence emission is collected from 530 to 700 nm. The excitation and emission slit are set at 3 nm.

Sensitivity Assay

1. Different concentrations of mRNA mimics are incubated with 100 nM H1-Tz and H2-PF in 100 μL 1× PBS buffer at 37 °C for 4 h.
2. Fluorescence spectra are then collected at room temperature. The excitation wavelength is 510 nm, and the fluorescence emission is collected from 530 to 700 nm. The excitation and emission slit are set at 3 nm.

Selectivity Assay

1. Different kinds of mRNA mimics (10 nM) are incubated with 100 nM H1-Tz and H2-PF in 100 μL 1× PBS buffer at 37 °C for 4 h.
2. Fluorescence spectra are then collected at room temperature. The excitation wavelength is 510 nm, and the fluorescence emission is collected from 530 to 700 nm. The excitation and emission slit are set at 3 nm.

Real-Time Fluorescence Response

1. Parameters are set before measurement: ex, 510 nm; emission, 545 nm; slit, 5 nm.
2. The instrument of Tecan Infinite M-1000 microplate reader is prewarmed up for 30 min to reach 37 °C.
3. Three samples are prepared as follows: (a) 100 nM Tz and 100 nM PF in 20 μL DMSO; (b) 100 nM H1-Tz and

100 nM H2-PF in 20 μL 1× PBS buffer; (c) 10 nM target mRNA mimics, 100 nM H1-Tz and 100 nM H2-PF in 20 μL 1× PBS buffer.

4. The reaction time is set as 10 h with a time interval of 20 min in a 384-well microplate.

3.6 Live Cell Culture and Confocal Imaging

1. HEK-293 T and HeLa cells are cultured in RPMI 1640 medium supplemented with 10% fetal bovine serum, 100 U/mL penicillin, and 100 μg/mL streptomycin. MCF-7 cells are cultured in DMEM medium supplemented with 10% fetal bovine serum, 100 U/mL penicillin, and 100 μg/mL streptomycin. Cells are all cultured at 37 °C in a humidified incubator containing 5% CO_2.
2. Cells are detached from culture flasks using 0.25% trypsin, seeded in confocal dishes, and further grown for 24 h to reach a 60–70% confluency. Cells are then washed three times with 1× PBS to remove the cell culture medium before transfection (*see* **Note 5**).
3. DNA transfections are performed using commercial reagents Lipofectamine 3000 according to the manufacturer's protocol with slight modification. Briefly, 250 μL of Opti-MEM supplemented with 2 μL lipofectamine 3000 are mixed with 250 μL Opti-MEM containing 5 μL P3000 reagents and DNA probes with a final concentration of 200 nM. The solution is then incubated at room temperature for 10 min.
4. Cells are then incubated with the prepared 500 μL transfection mixture at 37 °C for 6 h in a humidified incubator containing 5% CO_2.
5. After transfection, cells are washed three times with 1× PBS buffer and then stained with diluted solution of Hoechst 33342 (10 μg/mL) in the cell culture medium (500 μL) for 15 min at 37 °C in a humidified incubator containing 5% CO_2.
6. Finally, cells are washed with 1× PBS buffer and taken for confocal images using a Nikon microscopy. Fluorescence of Hoechst 33342 is obtained by excitation at 405 nm, and fluorescence of DNA probes is obtained by excitation at 488 nm.

4 Notes

1. The conjugation time could be optimized. Owing to the high click efficiency of Azide and DBCO, the reaction might be shortened.

2. The organic solvent should be removed out as much as possible, since a relative high percentage of DMSO might damage the C18 column.
3. Freeze-dried DNA products should be centrifuged for 5 min at 10000 rpm to make sure all DNA products will be located in the bottom of the tube before dissolution.
4. Load the gel solution carefully to avoid air bubble production. Additionally, both Acr-Bis and TEMED have neurotoxicity; you should handle with gloves and a mask.
5. It is better to add PBS solution slowly along the wall of the confocal dishes.

References

1. Chu HQ, Zhao J, Mi YS et al (2019) Near-infrared light-initiated hybridization chain reaction for spatially and temporally resolved signal amplification. Angew Chem Int Ed 58: 14877–14881
2. Choi HMT, Chang JY, Trinh LA et al (2010) Programmable in situ amplification for multiplexed imaging of mRNA expression. Nat Biotechnol 28:1208–1212
3. Chu YX, Deng AP, Wang WJ et al (2019) Concatenated catalytic hairpin assembly/hyperbranched hybridization chain reaction based enzyme-free signal amplification for the sensitive photoelectrochemical detection of human telomerase RNA. Anal Chem 91: 3619–3627
4. Liu L, Liu JW, Wu H et al (2018) Branched hybridization chain reaction circuit for ultrasensitive localizable imaging of mRNA in living cells. Anal Chem 90:1502–1505
5. Wu H, Chen TT, Wang XN et al (2020) RNA imaging in living mice enabled by an in vivo hybridization chain reaction circuit with a tripartite DNA probe. Chem Sci 11:62–69
6. Bai M, Chen F, Cao XW et al (2020) Intracellular entropy-driven multi-bit DNA computing for tumor progression discrimination. Angew Chem Int Ed 132:13369–13374
7. Li FY, Li G, Cao SJ et al (2021) Target-triggered entropy-driven amplification system-templated silver nanoclusters for multiplexed microRNA analysis. Biosens Bioelectron 172: 112757
8. Wang CY, Han CY, Du XX et al (2021) Versatile CRISPR-Cas12a-based biosensing platform modulated with programmable entropy-driven dynamic DNA networks. Anal Chem 93:12881–12888
9. Li HB, Han M, Weng X et al (2021) DNA-tetrahedral-nanostructure-based entropy-driven amplifier for high-performance Photoelectrochemical biosensing. ACS Nano 15:1710–1717
10. Yang TY, Fang J, Guo YC et al (2019) Fluorometric determination of microRNA by using an entropy-driven three-dimensional DNA walking machine based on a catalytic hairpin assembly reaction on polystyrene microspheres. Microchim Acta 186:574
11. Huang DJ, Cao T, Huang ZM et al (2019) Crosslinking catalytic hairpin assembly for high-contrast imaging of multiple mRNAs in living cells. Chem Commun 55:3899–3902
12. Karunanayake Mudiyanselage APKK, Yu QK, Leon-Duque MA et al (2018) Genetically encoded catalytic hairpin assembly for sensitive RNA imaging in live cells. J Am Chem Soc 140: 8739–8745
13. Zhang CH, Tang Y, Sheng YY et al (2016) Ultrasensitive detection of microRNAs using catalytic hairpin assembly coupled with enzymatic repairing amplification. Chem Commun 52:13584–13587
14. Wang JJ, Xia QF, Wu J et al (2021) A sensitive electrochemical method for rapid detection of dengue virus by CRISPR/Cas13a-assisted catalytic hairpin assembly. Anal Chim Acta 1187: 339131
15. Wei QM, Huang J, Jing L et al (2018) A DNA nanowire based localized catalytic hairpin assembly reaction for microRNA imaging in live cells. Chem Sci 9:7802–7808
16. Jia F, Wang DL, Lu XG et al (2018) Improving the enzymatic stability and the pharmacokinetics of oligonucleotides via DNA-backboned bottlebrush polymers. Nano Lett 18:7378–7382

17. Wu C, Cansiz S, Zhang L et al (2015) A non-enzymatic hairpin DNA cascade reaction provides high signal gain of mRNA imaging inside live cells. J Am Chem Soc 137:4900–4903

18. Zhao SJ, Zheng P, Wu Z et al (2022) DNA-templated bioorthogonal reactions via catalytic hairpin assembly for precise RNA imaging in live cells. Anal Chem 94:2693–2698

Chapter 15

Live-Cell Imaging of miRNAs with the Combination of CHA and HCR Techniques

Zai-Sheng Wu and Congcong Li

Abstract

The abnormal expression of miRNA-21 is closely related to many cancers, and its sensitive detection plays an important role in the diagnosis and treatment of cancers. In the report, we designed a non-enzymatic amplification sensing system based on the catalytic hairpin assembly (CHA) and palindrome-based hybridization chain reaction (PHCR) technique for the detection and imaging of miRNA in living cells. Based on PHCR technology, two ingeniously designed palindrome-contained fluorescently labeled hairpin probes are sequentially opened upon the stimulation of short oligonucleotide triggers, promoting the autonomous assembly of cross-linked network-like structures (CNSs) and generating fluorescence resonance energy transfer (FRET) signal. Moreover, the PHCR technique can be combined with CHA technology to sufficiently improve amplification efficiency and signal output. By utilizing the CHA-PHCR sensing system, one miRNA-21 activates many triggers through CHA process and in turn initiates the assembly of CNSs by PHCR reaction, efficiently amplifying the FRET signal output. As such, the sensing system can detect target miRNAs down to 10 pm with high detection specificity. Moreover, the CNS exhibits the enhanced intracellular stability, enabling the living cell imaging of miRNA-21. The CHA-PHCR sensing system can also screen the expression level of miRNA-21 in different living cells and differentiate cancer cells from healthy cells, which is consistent with the gold-standard PCR method.

Key words MiRNA detection, CHA-PHCR combination technique, Palindromic hairpin probe, Cross-linked network structures (CNSs), Live-cell imaging

1 Introduction

MiRNAs are a kind of endogenous, noncoding, single-stranded RNA with a length of 18 ~ 25 nucleotides, which are involved in the regulation of numerous biological processes, including cell proliferation, differentiation, and apoptosis [1–3]. Aberrant expression of some miRNAs has been reported to be associated with cancers [4]. Moreover, some miRNAs have been used as tumor markers for early clinical diagnosis of common cancers [5, 6].

Shin-ichi Sato (ed.), *Live-Cell RNA Imaging: Methods and Protocols*, Methods in Molecular Biology, vol. 2875,
https://doi.org/10.1007/978-1-0716-4248-1_15,

Hybridization chain reaction (HCR) proposed by Dirks and Pierce [7] is an efficient isothermal amplification technique and provides a ultra-high sensitivity. Moreover, the HCR process exhibits the sufficient design flexibility and does not require the participation of enzymes. Based on these advantages, HCR has a wide range of applications in biosensing and imaging. HCR products can enable biosensing and signal transduction by combining with functional moieties such as fluorophores [8, 9], nanoparticles (NPs) [10], and electrochemical reagents [11, 12].

Catalytic hairpin assembly (CHA), another enzyme-free, efficient isothermal amplification technique, was firstly proposed by Yin et al. [13]. As a simply designed universal isothermal nucleic acid amplification strategy, CHA has been used for the detection of a variety of biomarkers [14–16]. Moreover, CHA can be often combined with other isothermal amplification reactions to further improve the sensitivity and specificity of detection [17]. In addition, the palindrome-mediated cross-linking assembly is considered to be suitable for signal amplification in biosensing and bioimaging [18, 19].

In this contribution, we demonstrated a non-enzymatic palindrome-mediated assembly of DNA cross-linked network-like nanostructure for the amplification detection and imaging of miRNA-21 in living cells by combining CHA and HCR technologies (Scheme 1). The sensing system only contains four DNA probes: a couple of common hairpin probes (H1 and H2) and a couple of palindrome-incorporated fluorescently modified hairpin probes (f-PH1 and f-PH2). Under the stimulation of miRNA-21, the CHA process is initiated where H1 and H2 are involved, generating many DNA duplex (H1/H2) with sticky end and releasing target molecules. Each H1/H2 duplex can specifically trigger the subsequent palindrome-mediated HCR (PHCR) reaction, promoting the non-enzymatic self-assembly of f-PH1 and f-PH2 and thereby enabling fluorescence resonance energy transfer (FRET) signal output. The corresponding assembled product (abbreviated as dPCNS), cross-linked (C) network-like (N) DNA nanostructures (NS) based on dual non-enzymatic autonomous palindrome-mediated assembly process (dP), exhibits significantly enhanced biological stability, favoring the intracellular accumulation of fluorescence signal. During CHA process, the target is recycled but not consumed, and the PHCR reaction further

Scheme 1 (continued) (**a**) and living cells (**b**). For the CHA process, target can open H1 by hybridization, forming DNA duplex of Target/H1 with a long sticky end. Then, H2 replaces the target species through a toehold-mediated strand displacement reaction, forming a new DNA duplex structure (H1/H2) with a different sticky end. For the PHCR process, the exposed sticky end of H2 first hybridizes with f-PH1, releases its sticky end, and then opens the f-PH2, initiating the alternative hybridization between two partially complementary monomers in the longitudinal direction. The interaction between palindromic fragments can form the cross-linking structure in the lateral direction. This finally causes the generation of DNA cross-linked network structures. (Reproduced from Ref. [20] with permission from the American Chemical Society)

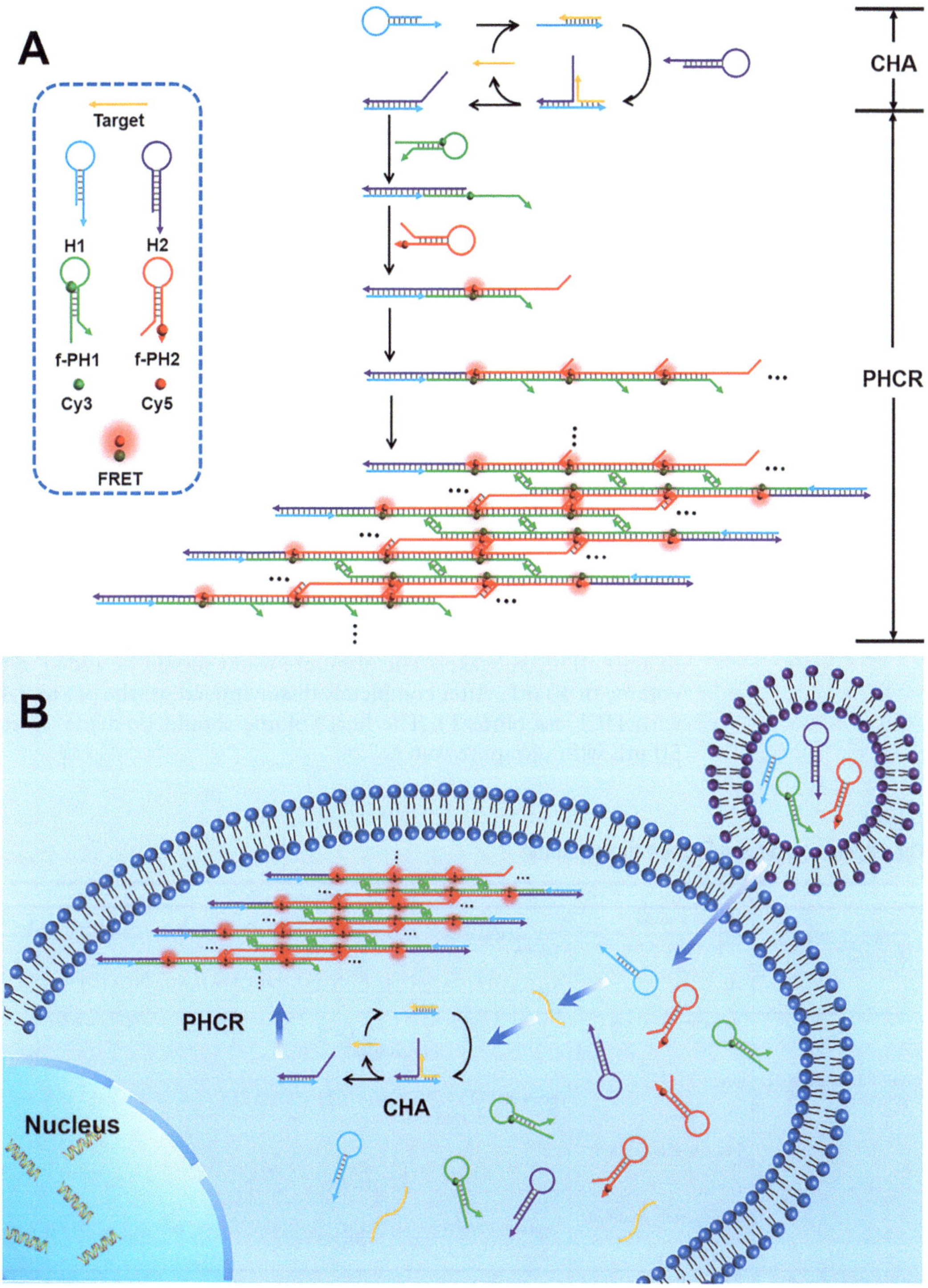

Scheme 1 Schematic illustration of nonenzymatic palindrome-mediated assembly of the cross-linked network structure (dPCNS) by double-amplification-cycle process upon target miRNA stimuli in a test tube

promotes the amplified output of the signal, forming the dual signal amplification effect and significantly improving the detection sensitivity. As such, the CHA-PHCR sensing system can quantitatively detect target miRNAs with LOD as low as 10 pM. Moreover, the expression levels of miRNAs in different cells can be specifically screened, and cancer cells can be precisely distinguished from healthy cells. In addition, the CHA-PHCR system can also be extended to detect other miRNAs by changing the target recognition sequence, showing excellent versatility.

2 Materials

All the aqueous solutions are prepared from ultrapure water purified by the Kerton lab MINI water purification system with resistivity ≥18.25 MΩ·Cm at 25 °C. All chemical reagents are of analytical grade. All solutions were stored at 4 °C before use. The oligonucleotides designed in this study are listed in Table 1.

2.1 CHA–PHCR-Based Assay

Reaction buffer (1× TAE/Mg^{2+} buffer): 20 mM Tris, 2 mM ethylenediaminetetraacetic acid (EDTA), and 12.5 mM $MgCl_2$, pH =7.4. Weigh 0.12 g Tris, 0.03 g EDTA, 0.06 g $MgCl_2$, and transfer to a 50 mL glass beaker. The ultrapure water should be added to a volume of 40 mL. After completely dissolving, adjust the pH to 7.4 with HCl (*see* **Note 1**). The final volume should be made up to 50 mL with ultrapure water.

Table 1
The oligonucleotides designed in the study[a]

Name		Sequences (5′-3′)
CHA for miRNA-21 detection	Hairpin strand 1 (H1)	ATCAGA**CTGATGTTGATACCTGCTCCATCC**TCAACATCAGTCTGATAAGCTA
	H2	**CTGATGTTGATACCTGCTCCATCC**TAGCTTATCAGACTGATGTTGAGGATGGAGCAGGTATCAACATCAGTCTGAT
PHCR for miRNA-21 detection	Palindromic hairpin 1 (PH1)	GGATGGAGCAGGTATCAACATCAGTCTGATGGTAGGATCAGACTGATGTTGATACCTGCTTTT**gatcgatc**
	PH2	**ccgtacgg**TTTATCAGACTGATGTTGATACCTGCTCCATCCAGCAGGTATCAACATCAGTCTGATCCTACC
	Fluorescently-modified PH1 (f-PH1)	GGATGGAGCAGGTATCAACATCAGTCTGATG/Cy3/GTAGGATCAGACTGATGTTGATACCTGCTTTT**gatcgatc**
	Fluorescently-modified PH2 (f-PH2)	**ccgtacgg**TTTATCAGACTGATGTTGATACCTGCTCCATCCAGCAGGTATCAACATCAGTCTGATCCTACC/Cy5/
DNA analogues of miRNA-21	Target-21	TAGCTTATCAGACTGATGTTGA

[a]The fragments with the same color have the same base sequence for one assay system for the detection of miRNA-21. The bold fragment in lowercase letters is the palindromic domain. The bold fragment in capital letters in both H1 and H2 is able to initiate PHCR by opening PH1 and f-PH1. The fragments with gray background in both H1 and H2 are completely complementary to miRNA-21 and its DNA analogue Target-21. The two underlined fragments in one probe indicate that they are complementary to each other, forming a hairpin structure. (Reproduced from Ref. [20] with permission from the American Chemical Society)

2.2 Native Polyacrylamide Gel Electrophoresis (nPAGE)

1. Resolving nPAGE buffer: 450 mM Tris, 450 mM boric acid, and 10 mM EDTA, pH =7.9. Weigh 54.5 g Tris, 27.8 g boric acid, and 2.9 g EDTA, and transfer to a 1 L glass beaker. The ultrapure water should be added to a volume of 900 mL. After the mixture is fully dissolved, the pH is adjusted to 7.9 with HCl (*see* **Note 1**). The final volume should be made up to 1 L with ultrapure water.
2. 10% ammonium persulfate: Dissolve 0.1 g ammonium persulfate in 1 mL of ultrapure water (*see* **Note 2**).
3. nPAGE running buffer: Mix 100 mL of resolving nPAGE buffer with 900 mL of ultrapure water.
4. 10 × SYBR Green I: Mix 1 μL of 10,000 × SYBR Green I with 999 μL of ultrapure water.

2.3 1% Agarose Gel Electrophoresis (AGE)

Resolving AGE buffer or AGE running buffer: Mix 100 mL of resolving nPAGE buffer with 900 mL of ultrapure water.

2.4 Cell Culture

Complete Dulbecco's modified Eagle's medium (C-DMEM): DMEM (high glucose) containing 10% fetal bovine serum (FBS) and 1% penicillin–streptomycin, which is prepared by mixing 89 mL of DMEM (high glucose) with 10 mL of fetal bovine serum (FBS) and 1 mL of penicillin–streptomycin.

3 Methods

Perform all experiments at room temperature unless otherwise noted.

3.1 CHA–PHCR-Based Assay

Dissolve all DNA sequences in 1× TAE/Mg^{2+} buffer (*see* **Note 3**). Add 2 μL of 1 μM H1, 2 μL of 1 μM H2, 2 μL of 10 μM f-PH1, 8 μL of 10 μM f-PH2, and 5.2 μL of 1× TAE/Mg^{2+} buffer into a 1.5-mL Eppendorf tube, obtaining the CHA–PHCR system.

Sample preparation for confocal imaging: Add Lipofectamine 3000 into the above-prepared CHA–PHCR system following the instruction manual, and incubate for 10 min, forming the Lip–CHA–PHCR complex.

3.2 Native Polyacrylamide Gel Electrophoresis (nPAGE)

1. Clean the glass plate, sample comb and spacer with ultrapure water in advance, and then wipe with ethanol and dried. Place the spacer between two cleaned glass plates and install the glass plates.
2. Add 2.8 mL of ultrapure water, 1.2 mL of resolving nPAGE buffer, 2.0 mL of 30% Acrylamide/Bis Solution (29:1), 0.1 mL 10% ammonium persulfate, and 5 μL of TEMED into a glass beaker (*see* **Note 4**). After mixing, add the resulting mixture to

the pre-installed glass plates immediately with a pipette. A 10-well sample comb should be then inserted in the middle of gel between the glass plates immediately (taking care not to introduce air bubbles). Once the gel is solidified, the sample comb is removed (*see* **Note 5**).

3. Mix 8 μL of DNA sample with 2 μL of 6 × loading buffer and 2 μL of 10 × SYBR Green I, and load into gel well (*see* **Note 6**). The 10% nPAGE is run on electrophoresis apparatus (BIO-RAD, USA) at 80 V in nPAGE running buffer for 70 min.
4. After electrophoresis, pry the two-layer gel glass plates apart by a spatula. Then gently rinse the gel on one of the glass plates with water, and transfer the gel to the hand.
5. The gel is placed in the ChemiDoc XRS$^+$ imaging system (BIO-RAD, USA) for imaging. Obtain the nPAGE image of CHA–PHCR system upon addition of Target-21 (DNA mimic of miRNA-21) (left panel of Fig. 1a).

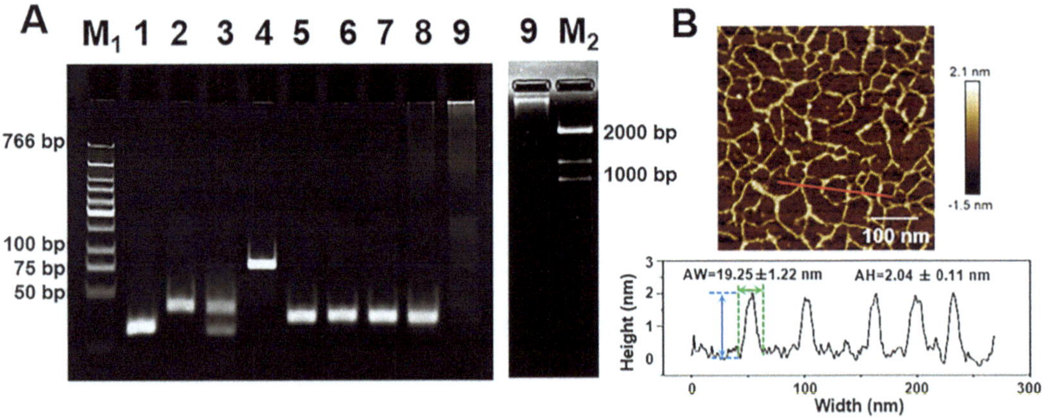

Fig. 1 Characterization of double-amplification-cycle-based palindrome-mediated cross-linked network structure (dPCNS). (**a**) Left panel: nPAGE analysis of the CHA–PHCR system in the presence of Target-21. Lane 1, 1 μM H1; lane 2, 1 μM H2; lane 3, 1 μM H1 + 1 μM H2; lane 4, 1 μM H1 + 1 μM H2 + 1 μM Target-21; lane 5, 1 μM PH1; lane 6, 1 μM PH2; lane 7, 1 μM PH1 + 1 μM PH2; lane 8, 100 nM H1 + 100 nM H2 + 1 μM PH1 + 4 μM PH2 (CHA-PHCR system); and lane 9, 100 nM H1 + 100 nM H2 + 1 μM PH1 + 4 μM PH2 + 40 nM Target-21. Right panel: 1% AGE analysis of lane 9. (**b**) AFM characterization of dPCNS formed in the CHA–PHCR system (final concentration, 50 nM H1, 50 nM H2, 0.5 μM PH1, and 2 μM PH2) in the presence of Target-21 (final concentration, 20 nM), accompanied by the cross-section profile along the red line. AW and AH indicate the average width and average height, respectively. (Reproduced from Ref. [20] with permission from the American Chemical Society)

3.3 1% Agarose Gel Electrophoresis (AGE)

1. Clean the mold and sample comb with ultrapure water in advance, then wipe with ethanol, and dry. Place the sample comb in the specified position.
2. Weigh 0.5 g agarose into a 250 mL conical flask, add 50 mL of resolving AGE buffer, shake well, and heat for 2 min in a microwave. Then, take out the conical flask. After the solution has cooled to a suitable temperature that does not burn the hands, pour the solution into the mold (taking care not to introduce air bubbles). Once the gel is solidified, the sample comb is removed (*see* **Note 7**).
3. Mix 8 μL of DNA sample with 2 μL of 6 × loading buffer and 2 μL of 10 × SYBR Green I, and load the resulting solution into gel well (*see* **Note 6**). The 1% AGE is run on electrophoresis apparatus (BIO-RAD, USA) at 80 V in AGE running buffer for 30 min.
4. After electrophoresis, the gel is placed in the ChemiDoc XRS$^+$ imaging system (BIO-RAD, USA) for imaging. Obtain the AGE result of CHA–PHCR system after addition of Target-21 (right panel of Fig. 1a).

3.4 Atomic Force Microscopy (AFM)

1. Fix the mica sheet on the round metal surface with double-sided tape.
2. Remove the top layers of mica with the help of a tape until obtaining satisfactorily smooth and clean mica surface.
3. Mix 10 μL of dPCNS products generated from CHA–PHCR system in the presence of Target-21 with 10 μL 1× TAE/Mg^{2+} buffer.
4. Drop 10 μL of resulting dPCNS products onto mica surface and let stand for 30 min. Wash the mica surface three times by 100 μL of ultrapure water, and blow dry with nitrogen (*see* **Note 8**).
5. Place the iron sheet on the object stage of MultiMode 8 atomic force microscope (Bruker, Germany) for AFM imaging. Obtain the AFM image of dPCNS products (Fig. 1b).

3.5 Fluorescence Detection

1. Detection feasibility of the CHA–PHCR system: Add 0.8 μL of 1 μM Target-21 or 0.8 μL of 1× TAE/Mg^{2+} buffer in 19.2 μL of CHA–PHCR system at 37 °C for 3 h. Then, the total volume should be supplemented to 200 μL with 1× TAE/Mg^{2+} buffer.
2. Performance target detection: Add 0.8 μL of Target-21 at a given concentration or 0.8 μL of 1× TAE/Mg^{2+} buffer into 19.2 μL of CHA–PHCR system, and incubate at 37 °C for 3 h. Then, the total volume should be supplemented to 200 μL with 1× TAE/Mg^{2+} buffer.

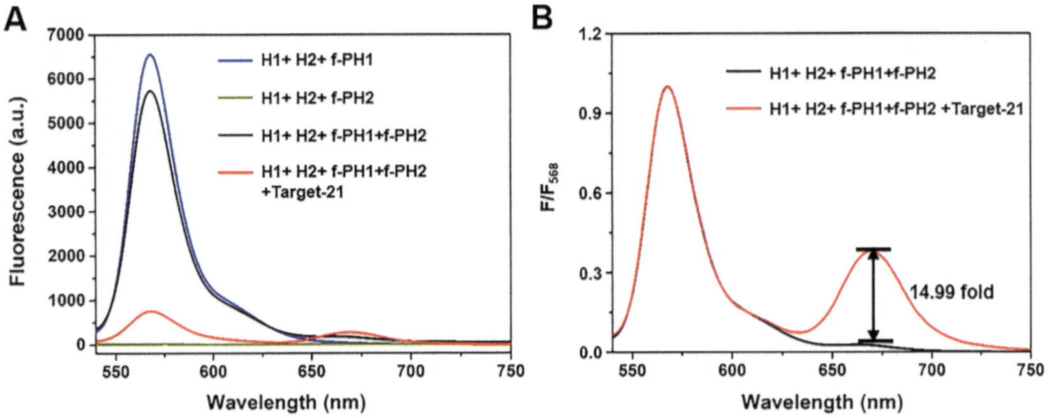

Fig. 2 Feasibility of the CHA–PHCR system for the amplification detection of miRNA-21. (**a**) Fluorescence spectra of the CHA–PHCR system in the presence of Target-21 and controls. The CHA–PHCR assay system in the absence and presence of Target-21 is (H1 + H2 + f-PH1 + f-PH2) and (H1 + H2 + f-PH1 + f-PH2+ Target-21), respectively. The final concentration of Target-21 is 4 nM. (**b**) Normalized fluorescence spectra corresponding to the samples described in panel A, which were obtained by dividing the emission FI at a given wavelength by FI at 568 nm (Cy3) (F568). The highest fluorescence ratio of target-to-noise (blank) found at 670 nm is defined as the FRET signal. (Reproduced from Ref. [20] with permission from the American Chemical Society)

3. Transfer the sample to the quartz fluorescence cell of F-7000 fluorescence spectrometer (Hitachi, Japan) for fluorescence detection. Fluorescence spectra upon addition of Target-21 (Figs. 2 and 3) (*see* **Note 9**).

3.6 Confocal Fluorescence Imaging

1. Culture human breast adenocarcinoma (MCF-7) cells, human cervical cancer (HeLa) cells, normal human hepatocytes (L02) cells, and human embryonic kidney 293 (HEK293) cells in cell culture dishes with C-DMEM medium.
2. When grown to a confluency of 70–80%, discard the C-DMEM.
3. Add 0.5 mL of 0.25% trypsin solution without EDTA to the cell culture dishes (*see* **Note 10**). Generally, the digestion time is set to 1–3 min.
4. When there are gaps with the size of pinholes in the cell layer adhered to bottle wall, the trypsin solution is discarded, and 1 mL of C-DMEM is added to terminate the digestion.
5. Repeatedly pipet the cells with C-DMEM in the culture flask until the cells are completely detached.
6. Place a correct-size coverslip in a 24-well plate. Then, the appropriate number of digested cells is plated in the 24-well plate. After the cells adhere to the wall, C-DMEM is added.

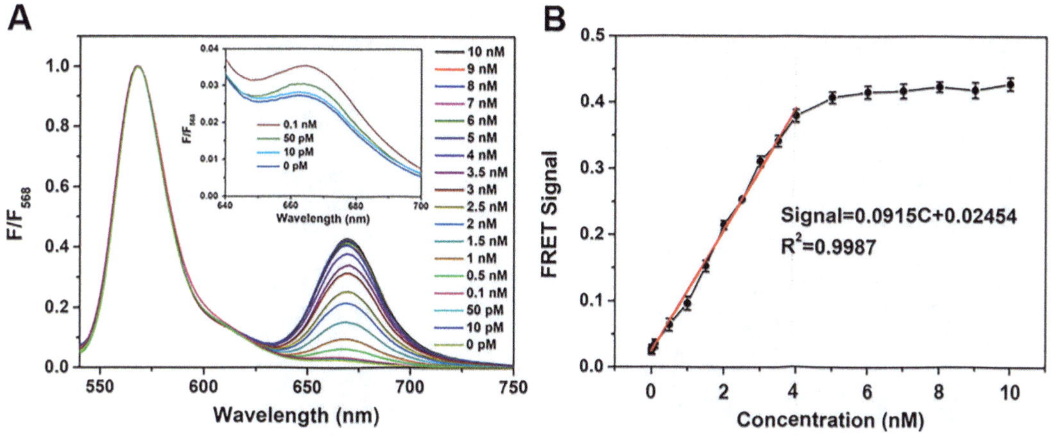

Fig. 3 Assay performance of the CHA–PHCR system. (**a**) Normalized fluorescence spectrum of the CHA–PHCR system in the presence of different concentrations of Target-21. (**b**) Dynamic response relationship between the FRET signal and Target-21 concentration, ranging from 0 to 10 nM, accompanied by a linear regression equation. The CHA–PHCR assay system contains H1 (10 nM), H2 (10 nM), f-PH1 (100 nM), and f-PH2 (400 nM). Error bars represent standard deviations from three independent experiments. (Reproduced from Ref. [20] with permission from the American Chemical Society)

7. When cells are grown to 70–80% confluence, discard the original C-DMEM. Wash the cells three times with 400 μL of PBS buffer. Add Lip–CHA–PHCR complex into the cells, and incubate for 4 h at cell incubator (*see* **Note 9**).
8. Remove the Lip–CHA–PHCR complex from 24-well plate. Wash the cells three times with 400 μL of PBS buffer. Add 400 μL of 12.5 μg/mL Hoechst 33342 to 24-well plate, and incubate for 10 min. Discard Hoechst 33342, and wash cells three times with 400 μL of PBS buffer (*see* **Note 9**).
9. Drop 1 μL of Antifade Mounting Medium onto glass slide. Remove the coverslip in the 24-well plate, and place it upside down on the glass slide with Antifade Mounting Medium for confocal imaging (Fig. 4) (*see* **Note 9**).

4 Notes

1. When adjusting the pH value, it is necessary to add dropwise the concentrated HCl several times to avoid the sudden dropping of pH value below the required pH value.
2. The 10% ammonium persulfate can be stored at 4 °C for 15 days.
3. For dissolving DNA sequences, centrifuge the DNA powder to the bottom of tubes, open the cap carefully and gently, and add 1× TAE/Mg^{2+} buffer. Then, place it on a vortex shaker to mix thoroughly before use.

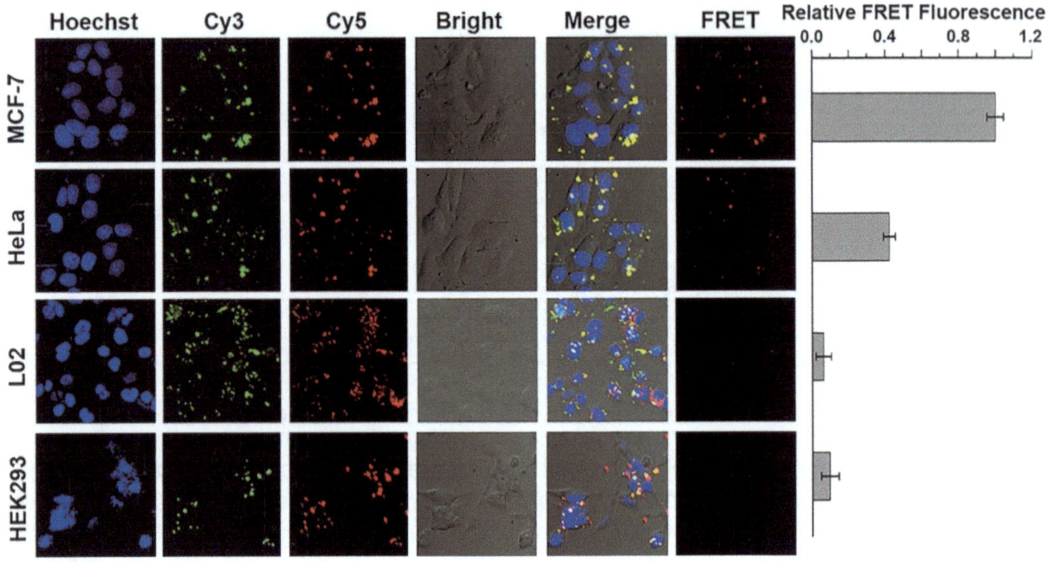

Fig. 4 Confocal imaging of different cell lines with different miRNA-21 expression levels. The quantitative analysis of relative FRET signal intensity is shown in the right panel. The CHA–PHCR system consists of H1 (10 nM), H2 (10 nM), f-PH1 (100 nM), and f-PH2 (400 nM) in the final reaction solution. The MCF-7, HeLa, L02, and HEK293 cells were separately incubated in the CHA–PHCR system for 4 h. Error bars represent standard deviations from three independent experiments. (Reproduced from Ref. [20] with permission from the American Chemical Society)

4. Both 30% Acrylamide/Bis Solution (29:1) and TEMED are highly toxic. Wear protective clothing, gloves, and masks when operating. Moreover, operate in a fume hood.
5. If the prepared gel is not used immediately, it should be sealed with plastic wrap and stored at 4 °C. Moreover, use the gel within 12 h because the gel will dry out if stored for too long.
6. Load the sample into the gel wells slowly to prevent spillage of samples into the running buffer.
7. When the gel solution is poured into the mold, the mold should be placed on a flat table; otherwise the thickness of the solidified agarose gel will be not uniform. In addition, the prepared AGE can be stored in resolving AGE buffer for 15 days at room temperature.
8. When cleaning mica sheets, be careful not to let water droplets touch the surrounding iron sheets to prevent the contamination by impurities on the iron sheets.
9. The experimental process needs to be carried out in the dark.
10. The volume of trypsin solution can be changed depending on the amount of cells in order to make sure that all the cells can be fully immersed.

Acknowledgments

This work was supported by the National Natural Science Foundation of China (NSFC) (grant no 22174020).

References

1. Cheloufi S, Dos Santos CO, Chong MM, Hannon GJ (2010) A dicer-independent miRNA biogenesis pathway that requires ago catalysis. Nature 465(7298):584–589
2. Ong S-G, Lee WH, Kodo K, Wu JC (2015) MicroRNA-mediated regulation of differentiation and trans-differentiation in stem cells. Adv Drug Deliv Rev 88:3–15
3. Shimono Y, Zabala M, Cho RW, Lobo N, Dalerba P, Qian D et al (2009) Downregulation of miRNA-200c links breast cancer stem cells with normal stem cells. Cell 138(3): 592–603
4. De Leeuw D, Verhagen H, Denkers F, Kavelaars F, Valk P, Schuurhuis GJ et al (2016) MicroRNA-551b is highly expressed in hematopoietic stem cells and a biomarker for relapse and poor prognosis in acute myeloid leukemia. Leukemia 30(3):742–746
5. Adams BD, Anastasiadou E, Esteller M, He L, Slack FJ (2015) The inescapable influence of noncoding RNAs in cancer. Cancer Res 75(24):5206–5210
6. Zhang J, Xiao X, Liu J (2015) The role of circulating miRNAs in multiple myeloma. Sci China Life Sci 58(12):1262–1269
7. Dirks RM, Pierce NA (2004) Triggered amplification by hybridization chain reaction. Proc Natl Acad Sci 101(43):15275–15278
8. Bi S, Chen M, Jia X, Dong Y, Wang Z (2015) Hyperbranched hybridization chain reaction for triggered signal amplification and concatenated logic circuits. Angew Chem Int Ed 54(28):8144–8148
9. Xuan F, Hsing I-M (2014) Triggering hairpin-free chain-branching growth of fluorescent DNA dendrimers for nonlinear hybridization chain reaction. J Am Chem Soc 136(28): 9810–9813
10. Liu P, Yang X, Sun S, Wang Q, Wang K, Huang J et al (2013) Enzyme-free colorimetric detection of DNA by using gold nanoparticles and hybridization chain reaction amplification. Anal Chem 85(16):7689–7695
11. Hou T, Li W, Liu X, Li F (2015) Label-free and enzyme-free homogeneous electrochemical biosensing strategy based on hybridization chain reaction: a facile, sensitive, and highly specific microRNA assay. Anal Chem 87(22): 11368–11374
12. Zhang B, Liu B, Tang D, Niessner R, Chen G, Knopp D (2012) DNA-based hybridization chain reaction for amplified bioelectronic signal and ultrasensitive detection of proteins. Anal Chem 84(12):5392–5399
13. Yin P, Choi HM, Calvert CR, Pierce NA (2008) Programming biomolecular self-assembly pathways. Nature 451(7176): 318–322
14. Chen RP, Blackstock D, Sun Q, Chen W (2018) Dynamic protein assembly by programmable DNA strand displacement. Nat Chem 10(4):474–481
15. Park C, Park H, Lee HJ, Lee HS, Park KH, Choi C-H et al (2019) Double amplified colorimetric detection of DNA using gold nanoparticles, enzymes and a catalytic hairpin assembly. Microchim Acta 186(1):1–8
16. Zhao R-N, Feng Z, Zhao Y-N, Jia L-P, Ma R-N, Zhang W et al (2019) A sensitive electrochemical aptasensor for mucin 1 detection based on catalytic hairpin assembly coupled with PtPdNPs peroxidase-like activity. Talanta 200:503–510
17. Yan Y, Shen B, Wang H, Sun X, Cheng W, Zhao H et al (2015) A novel and versatile nanomachine for ultrasensitive and specific detection of microRNAs based on molecular beacon initiated strand displacement amplification coupled with catalytic hairpin assembly with DNAzyme formation. Analyst 140(16): 5469–5474
18. Liu R, Zhang S, Zheng T-T, Chen Y-R, Wu J-T, Wu Z-S (2020) Intracellular nonenzymatic in situ growth of three-dimensional DNA nanostructures for imaging specific biomolecules in living cells. ACS Nano 14(8): 9572–9584
19. Zeng R, Zhang L, Luo Z, Tang D (2019) Palindromic fragment-mediated single-chain amplification: an innovative mode for photoelectrochemical bioassay. Anal Chem 91(12): 7835–7841
20. Li C, Zhang J, Gao Y, Luo S, Wu Z-S (2022) Nonenzymatic autonomous assembly of cross-linked network structures from only two palindromic DNA components for intracellular fluorescence imaging of miRNAs. ACS Sensors 7(2):601–611

Chapter 16

Live-Cell Imaging of Multiple Endogenous mRNAs Using RNA Aptamers and Chemical Probes

Kenji Yatsuzuka, Yousuke Katsuda, and Shin-ichi Sato

Abstract

Imaging of RNA dynamics in living cells is increasingly important to understanding and measuring spatially restricted gene expression. We recently developed a live-cell RNA imaging method that combines an RNA aptamer with a fluorescent chemical probe. The method uses a combination of transfection of a plasmid encoding a gene-specific RNA aptamer with a cell-permeable synthetic small molecule whose fluorescence is restored only when the RNA aptamer hybridizes with its cognitive mRNAs. The versatile method permits the observation of the formation process of stress RNA granules crucial for cellular response to the environment. Simple approaches for simultaneous imaging of multiple RNAs would be essential to gain deeper insights into the functions and dynamics of RNA in cells.

Key words Live-cell imaging, mRNA, RNA aptamer, Small molecule, Chemical biology

1 Introduction

Imaging of RNAs in living cells is a powerful approach to understanding the intracellular dynamics of RNAs and to measuring spatiotemporal gene expression. A number of different methods have been developed for detecting RNA targets [1–15]. The use of molecular beacons, the fluorescence of which changes upon binding to native mRNA targets, is known to be one of the best approaches [16–19]. The high designability for specific RNA targets permits visualization of a wide range of endogenous RNAs. However, several challenges remain, for example, low stability of the beacons in cells produces relatively high background fluorescence. A new technology for RNA imaging is needed to overcome such drawbacks. We recently developed a convenient and versatile method that permits spatiotemporal imaging of specific native RNAs or proteins in living cells [20–23]. The method combines a tailor-made RNA aptamer with a cell-permeable, quenched fluorescent chemical probe. In our method, the chemical probe is

Shin-ichi Sato (ed.), *Live-Cell RNA Imaging: Methods and Protocols*, Methods in Molecular Biology, vol. 2875,
https://doi.org/10.1007/978-1-0716-4248-1_16,

conjugated with fluorescent quenchers BHQ1 or DNB, and their quenched fluorescence is restored when the quencher segment binds to the RNA aptamer only in the presence of the mRNA of interest. When the RNA aptamer, which is encoded by a transfected plasmid, hybridizes to a complementary sequence in the target mRNA, it forms a binding loop for BHQ1 or DNB to restore the fluorescence of the conjugate. This versatile method permits visualization of non-engineered, endogenous mRNAs in living cells. Here, we describe the synthesis of fluorescent probes, the selection of RNA aptamers which bind to the quenchers, and the construction of RNA-targeting aptamers that hybridize to the target RNAs. We also describe here how to use our imaging technology for multiple RNA imaging in mammalian living cells. This RNA imaging technology could provide new insights into RNA functions and gene expression.

2 Materials

All solutions are prepared using ultrapure water and analytical grade reagents. Reagents are prepared and stored at room temperature unless otherwise indicated. Distilled water and special grade solvents and reagents are used for chemical syntheses.

2.1 Reagents

1. EAH Sepharose™ 4B resin.
2. BHQ1 carboxylic acid (*see* **Note 1**).
3. BHQ1 amine (*see* **Note 1**).
4. 4-(4,6-Dimethoxy-1,3,5-triazin-2-yl)-4-methylmorpholinium chloride n-hydrate (DMT-MM).
5. 2,4-Dinitrofluorobenzene.
6. *N*-(*tert*-butoxycarbonyl)-2,2′-(ethylenedioxy)diethylamine (BocNH-PEG2-NH_2).
7. Diglycolic anhydride.
8. *N*,*N*-diisopropylethylamine.
9. *N*-hydroxysuccinimide (HO-Su).
10. 1-(3-Dimethylaminopropyl)-3-ethylcarbodiimide hydrochloride (WSCI·HCl, EDC·HCl).
11. NH_4Cl.
12. Dimethyl sulfoxide (DMSO).
13. Ethyl acetate (AcOEt).
14. Triethylamine.
15. Acetic acid.
16. Dioxane.

17. Methanol.
18. *N,N*-dimethylformamide (DMF).
19. Dichloromethane (CH_2Cl_2).
20. Methoxytrityl-N-dPEG8-TFP ester.
21. Cy5-NHS-ester (*see* **Note 1**).
22. *O*-(2-aminoethyl)-*O′*-[2-(Boc-amino)ethyl]hexaethylene glycol (BocNH-PEG7-NH_2).
23. deSO_3-DY520-XL (*see* **Notes 1** and **2**).
24. 1-((Dimethylamino)(dimethyliminio)methyl)-1H-[1,2,3]triazolo[4,5-b]pyridine 3-oxide hexafluorophosphate (HATU).
25. 3H-[1,2,3]triazolo[4,5-b]pyridin-3-ol (HOAt).
26. Trifluoroacetic acid (TFA).
27. PCR Master mix.
28. 3 M sodium acetate buffer solution (pH 5.2).
29. 70% ethanol.
30. TE buffer solution (pH 8.0).
31. T7 in vitro transcription kit including DNase (MEGAscript® T7 Kit).
32. 10 M ammonium acetate solution (pH 7.0).
33. DNA purification columns.
34. 0.22 μm centrifugal filter units.
35. Reverse transcription (RT) kit.
36. *Eco*R I, *Bam*H I, *Bgl* II, and *Hind* III restriction enzymes.
37. DNA ligation kit.
38. Competent high DH5α.
39. Miniprep Kit.
40. Midiprep Kit.
41. 4% paraformaldehyde (PFA) in 0.1 M PBS.
42. pUC19 DNA.
43. pSuper.neo vector.
44. 1× PBS.
45. Hoechst 33342.
46. Sodium arsenite.

2.2 Oligonucleotides

1. N60 template DNA: 5′-GAA TTC CGC GTG TGC ACA CC-N60-GTC CGT TGG GAT CCT CAT GG-3′.
2. Forward primer: 5′-GCT AAT ACG ACT CAC TAT AGG GAA TTC CGC GTG TGC ACA CC-3′ (T7 promoter sequence is underlined).

3. Reverse primer: 5′-CCA TGA GGA TCC GAA CGG AG-3′.
4. RT aptamer template 1: 5′-GAT CCC CGG AGC AAT GAT GGC **CTA GAT AAA TTC GGA** GCT TGA TCT TCA TTT TTA-3′ (RNA recognition arms are underlined, and the BHQ1-binding loop is bold).
5. RT aptamer template 2: 5′-AGC TTA AAA ATG AAG ATC AAG CTC CGA ATT TAT CTA GGC CAT CAT TGC TCC GGG −3′.

2.3 Buffers and Reaction Mixes

1. PCR mixture: 10 μL N60 template DNA, 250 μL 2× KOD FX Neo buffer, 100 μL dNTP mixture (2 mM each), 10 μL 10 μM Forward primer, 10 μL 10 μM reverse primer, 10 μL 1.0 U/μL KOD FX Neo, 110 μL RNase-free water.
2. In vitro transcription mixture: 8 μL template DNA, 2 μL 10× T7 reaction buffer, 2 μL 2 mM ATP, 2 μL 2 mM CTP, 2 μL 2 mM GTP, 2 μL 2 mM UTP, 2 μL T7 enzyme mix.
3. RT reaction mixture: 7 μL template RNA, 0.5 μL 10 μM reverse primer, 2 μL 5× RT reaction buffer, 0.5 μL RT enzyme mix.
4. Ligation mixture: 5 ng insert DNA, 5 ng linearized plasmid DNA, RNAse-free water up to 1 μL, 1 μL ligation mix.
5. Annealing buffer: 10 mM Tris–HCl, pH 7.6, 100 mM KCl, and 5 mM $MgCl_2$.
6. 2× SSC buffer: 300 mM NaCl, 30 mM sodium citrate, pH 7.0 50 v/v % formamide.
7. Pre-hybridization buffer: 300 mM NaCl, 30 mM sodium citrate, pH 7.0, 10 w/v % dextran sulfate, 2 mM vanadyl-ribonucleoside complex, 0.02 w/v % RNase-free BSA, 40 μg *E. coli* tRNA, and 30 v/v % formamide.

2.4 Cell Culture

Cells are maintained in Dulbecco's modified Eagle medium, supplemented with 100 units/mL penicillin, 100 μg/mL streptomycin sulfate, and 10% (v/v) fetal bovine serum, at 37 °C in a humidified 5% CO_2 incubator.

1. Trypsin-EDTA (0.25%), phenol red.
2. Transfection reagent.
3. PDL-coated glass-bottom ViewPlate-96.

2.5 Equipment

1. HPLC system.
2. 5 μm, 20.0 × 100 mm preparative columns.
3. Fluorescence spectrometer.
4. Spectrophotometer.
5. Tabletop high-speed micro centrifuge.

6. Thermal cycler.
7. Confocal microscope.
8. LC-MS system.
9. 2.2 μm, 3 × 50 mm analytical columns.
10. Freeze dryer.

3 Methods

3.1 Preparation of the BHQ1-Immobilized Resin

Synthesis of the BHQ1-immbilized resin is summarized in Scheme 1a.

1. Place 0.5 mL of EAH Sepharose™ 4B resin into a Poly-Prep Chromatography Column, and allow it to drain by gravity.
2. Wash the resin thoroughly with at least three bed volumes of dioxane-H_2O (3:1, v/v).
3. Add 1.0 μmol BHQ1 carboxylic acid, 0.10 mmol triethylamine, and 0.10 mmol DMT-MM in 0.5 mL of dioxane-H_2O (3:1, v/v) to the drained resin, and shake the suspension at room temperature overnight (*see* **Note 3**).
4. Drain the coupling solution from the resin, and add 0.50 mmol acetic acid and 0.50 mmol DMT-MM freshly prepared in 0.5 mL of dioxane-H_2O (3:1, v/v) to cap unreacted amino groups on the resin (*see* **Note 3**).

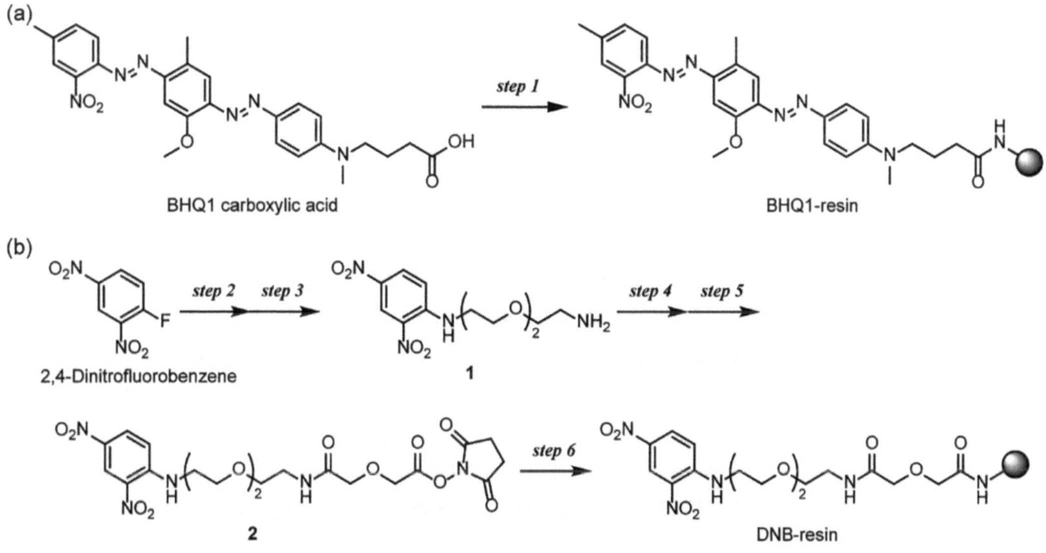

Scheme 1 Synthesis of BHQ1-immobilized resin (**a**) and DNB-immobilized resin (**b**)

5. After overnight incubation, drain the resin, and wash thoroughly with dioxane-H_2O (1:1, v/v) followed by methanol-H_2O (1:1, v/v).
6. Re-suspend the resin with methanol-H_2O (1:3, v/v) to make a 50% slurry. Store at 4 °C until use.

3.2 Preparation of the DNB-Immobilized Resin

Synthesis of the DNB-immobilized resin is summarized in Scheme 1b.

1. Add 0.85 mmol 2,4-dinitrofluorobenzen to a solution of 0.94 mmol BocNH-PEG2-NH_2 and 1.7 mmol *N,N*-diisopropylethylamine in 5 mL of CH_2Cl_2.
2. Stir the mixture for 2 h at room temperature, and add 0.2 M HCl aq. to the mixture (*see* **Note 4**).
3. Extract the products with CH_2Cl_2 (10 mL × 3).
4. After washing the products with 10 mL of brine, remove the solvent.
5. Purify the residue by silica gel column chromatography using hexane/AcOEt (77:23 to 19:81) as eluent.
6. Add 5 mL of 4 M HCl/AcOEt to the purified products.
7. Stir the solution for 3 h at room temperature (*see* **Note 4**).
8. Collect the resulting solid by filtration to obtain compound **1** as an HCl salt.
9. Add 0.17 mmol *N,N*-diisopropylethylamine to a solution of 0.14 mmol compound **1** and 0.16 mmol diglycolic acid in 1 mL of CH_2Cl_2.
10. Stir the mixture for 2 h at room temperature, and add 1 M HCl aq. to the mixture (*see* **Note 4**).
11. Extract the products with CH_2Cl_2 (10 mL × 3).
12. After washing the products with 10 mL of brine, remove the solvent.
13. Add a solution of 0.17 mmol WSCI·HCl and 0.17 mmol HO-Su in 1 mL of CH_2Cl_2 to the residue.
14. Stir the mixture for 3 h at room temperature, and add 10 mL of H_2O to the mixture (*see* **Note 4**).
15. Extract the products with CH_2Cl_2 (15 mL × 2).
16. After washing the products with 10 mL of sat. NH_4Cl aq. and 10 mL of brine, remove the solvent to obtain compound **2**.
17. Place 2 mL of EAH Sepharose™ 4B resin into a Poly-Prep Chromatography Column, and allow it to drain by gravity.
18. Wash the resin thoroughly with at least three bed volumes of dioxane-H_2O (3:1, v/v).

19. Add the compound **2**, 42 μmol *N,N*-diisopropylethylamine in 1 mL of dioxane-H_2O (3:1, v/v) to the drained resin, and shake the suspension at room temperature overnight.
20. Drain the coupling solution from the resin, and add 0.50 mmol acetic acid and 0.50 mmol DMT-MM freshly prepared in 0.5 mL of dioxane-H_2O (3:1, v/v) to cap unreacted amino groups on the resin (*see* **Note 3**).
21. After overnight incubation, drain the resin, and wash thoroughly with dioxane-H_2O (1:1, v/v) followed by methanol-H_2O (1:1, v/v).
22. Re-suspend the resin with methanol-H_2O (1:3, v/v) to make a 50% slurry. Store at 4 °C until use.

3.3 Preparation of BHQ1-Cy5 Probes

Synthesis of the BHQ1-Cy5 probe is summarized in Scheme 2a.

1. Add 15 μmol methoxytrityl-N-dPEG8-TFP ester to a solution of 10 μmol BHQ1 amine and 40 μmol *N,N*-diisopropylethylamine in 50 μL of DMSO.
2. Stir the mixture for 3 h at room temperature (*see* **Note 5**).
3. Add 300 μL of H_2O and 500 μL of hexane to the mixture.

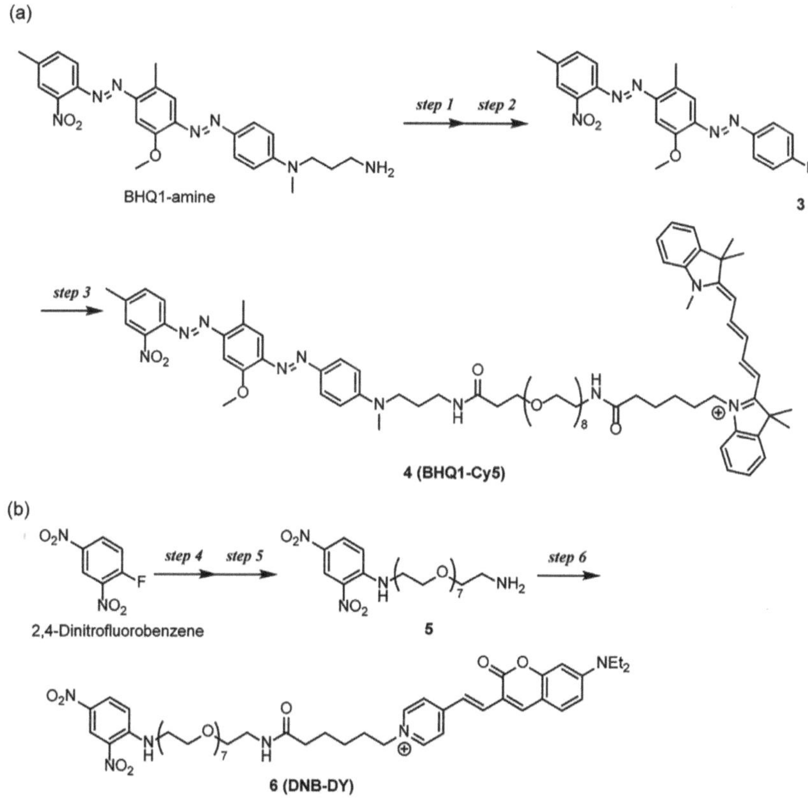

Scheme 2 Synthesis of BHQ1-Cy5 (**a**) and DNB-DY (**b**) probes

4. Collect the resulting solid by centrifugation (21,500× *g* for 10 min).
5. Wash the solid with H_2O (500 μL) and dry it in vacuo.
6. Dissolve the residue in 400 μL of TFA.
7. Stir the mixture for 2 h at room temperature (*see* **Note 5**).
8. Remove the solvent.
9. Purify the crude product by preparative HPLC (0–3 min, CH_3CN, 55%; 3–23 min, CH_3CN, 55–65% at RT, $t_R = 20.6$ min.) to obtain compound **3**.
10. Add 0.6 μmol Cy5-NHS-ester to a solution of 0.5 μmol compound **3** and 2 μmol *N,N*-diisopropylethylamine in 20 μL of DMSO.
11. Stir the mixture for 12 h at room temperature (*see* **Note 5**).
12. Purify the crude mixture by preparative HPLC (0–3 min, CH_3CN, 55%; 3–33 min, CH_3CN, 55–80% at RT, $t_R = 29.1$ min.) to obtain compound **4** (BHQ1-Cy5 probe) (*see* **Note 6**).
13. The BHQ1-Cy5 probe is dissolved in DMSO and stored at − 30 °C.

3.4 Preparation of DNB-DY Probes

Synthesis of the DNB-DY probe is summarized in Scheme 2b.

1. Add 5.0 μmol 2,4-dinitrofluorobenzen to a solution of 6.0 μmol BocNH-PEG7-NH_2 and 10 μmol *N,N*-diisopropylethylamine in 50 μL of CH_3CN.
2. Stir the mixture for 2 h at room temperature, and add 0.2 M HCl aq. to the mixture (*see* **Note 5**).
3. Extract the products with CH_2Cl_2 (200 μL × 4).
4. After washing the products with 100 μL of brine, remove the solvent.
5. Dissolve the residue in 300 μL of 4 M HCl/AOEt and 100 μL of AcOEt.
6. Stir the mixture for 2 h at room temperature (*see* **Note 5**).
7. Added 2 mL of hexane to the mixture.
8. Collect the resulting solid by centrifugation (21,500× *g* for 10 min).
9. Wash the solid with hexane (500 μL) and dry it in vacuo to obtain compound **5** as a HCl salt.
10. Add 1 μmol *N,N*-diisopropylethylamine to a solution of 0.25 μmol compound **5**, 0.3 μmol deSO_3-DY520-XL, 0.5 μmol HATU, and 0.5 μmol HOAt in 20 μL of DMSO.
11. Stir the mixture for 16.5 h at room temperature (*see* **Note 5**).

12. Purify the crude mixture by preparative HPLC (0–3 min, CH_3CN, 50%; 3–23 min, CH_3CN, 50–60% at RT, $t_R = 15.7$ min.) to obtain compound **6** (DNB-DY probe) (*see* **Note 6**).
13. The DNB-DY probe is dissolved in DMSO and stored at −30 ° C.

3.5 In Vitro Selection of RNA Aptamers

In vitro selection is performed according to standard procedure for isolating RNA aptamers for BHQ1 and DNB [24–26] (Fig. 1). The selection cycle is repeated several times to enrich RNA species specifically bound to BHQ1 or DNB.

3.5.1 Preparation of DNA Pools (See Note 7)

1. Prepare the PCR mixture in PCR tubes.
2. Place the tubes in a thermal cycler, and perform the PCR amplification using the following program: 2-min initial denaturation at 94 °C, followed by 30 sequential cycles of denaturation at 98 °C for 10 s, annealing at 55 °C for 30 s, and extension at 68 °C for 30 s.
3. Precipitate the DNA by adding a one-tenth volume of 3 M sodium acetate and a 1.1 volume of isopropanol. Centrifuge the sample at 15,000× *g* for 10 min to pellet the DNA.
4. Remove the supernatant carefully, and rinse the pellet with 70% ethanol.
5. Air-dry the pellet for 5 min, and dissolve it in a 20-μL TE buffer solution.

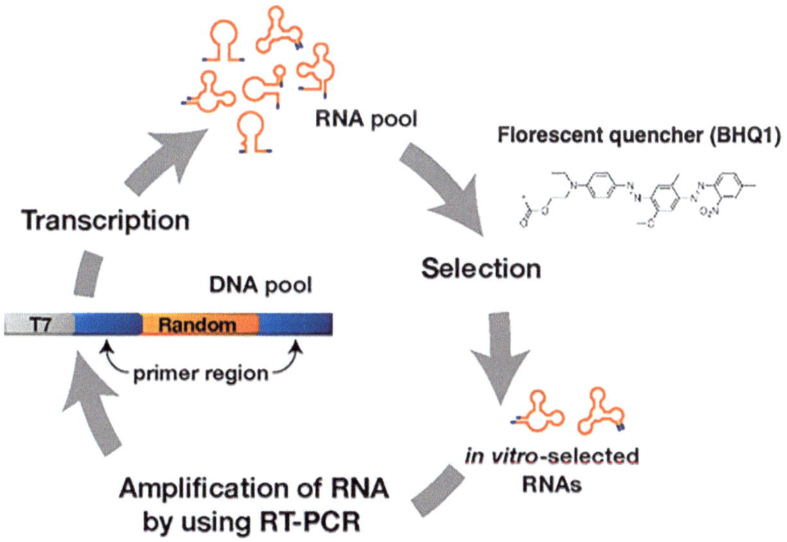

Fig. 1 In vitro selection of BHQ1 aptamers

3.5.2 Preparation of RNA Pools

1. Prepare the in vitro transcription reaction at room temperature in the order as described in the Materials.
2. Incubate the reaction mixture overnight at 37 °C.
3. Add 1 μL of DNase and incubate for 15 min at 37 °C.
4. Precipitate the RNA by adding a one-third volume of 10 M ammonium acetate and a 1.5 volume of isopropanol. Centrifuge the sample at 15,000× *g* for 10 min at 4 °C to pellet the RNA.
5. Thoroughly remove the supernatant, and dissolve the pellet in 0.5 mL of RNase-free water.
6. Apply the sample to the pre-equilibrated NAP-5 column, and elute the sample with 1 mL of RNase-free water (*see* **Note 8**).
7. Precipitate the RNA by adding a one-tenth volume of 3 M sodium acetate and a 1.1 volume of isopropanol. Centrifuge the sample at 15,000× *g* for 10 min to pellet the RNA.
8. Remove the supernatant, and rinse the pellet with 70% ethanol.
9. Air-dry the pellet for 5 min, and dissolve it in 100 μL of the annealing buffer.
10. Use the solution as an RNA pool for in vitro selection.

3.5.3 In Vitro Selection of RNA Aptamer to BHQ1- or DNB-Immobilized Resin (See Note 9)

1. Place 20 μL of the BHQ1- or DNB-immobilized resin (50% slurry) in an empty centrifugal filter unit, and wash the resin three times with 400 μL of the annealing buffer.
2. Apply the RNA pool solution to the resin, and incubate on ice for 30 min with manual shaking every 3 min.
3. Drain and wash the resin with annealing buffer to remove unbound RNAs.
4. Elute the bound RNAs three times with 130 μL of the annealing buffer saturated with free BHQ1 amine or compound **1**.
5. Precipitate the RNAs by adding a one-tenth volume of 3 M sodium acetate and a 1.1 volume of isopropanol. Centrifuge the sample at 15,000× *g* for 10 min to pellet the RNA.
6. Remove the supernatant and rinse the pellet with 70% ethanol.
7. Air-dry the pellet for 5 min, and dissolve it in a 15 μL TE buffer solution.

3.5.4 Reverse Transcription

1. Heat the eluted resin-bound RNA for 5 min at 65 °C, and then immediately chill it on ice.
2. Prepare the RT reaction mixture.
3. The reaction is performed for 15 min at 37 °C, followed by heat inactivation of RTase at 98 °C for 5 min.

4. 4.6 μL of the reverse-transcribed products are used as templates for PCR amplification to generate the DNA pool for the next round of selection (*see* **Note 10**).

3.5.5 Cloning and Sequencing of RNA Aptamers for BHQ1

1. After the final round of selection, digest the resulting DNA pool with restriction enzymes *Eco*R I and *Bam*H I.
2. Prepare the ligation mixture with *Eco*R I and *Bam*H I pUC19 plasmid as a vector.
3. Incubate the mixture for 30 min at 16 °C to allow the ligation reaction.
4. Transform DH5α with the ligation product, and heat shock at 42 °C for 1 min, and spread the cells on an LB plate containing 100 μg/mL of carbenicillin.
5. Incubate the plate overnight at 37 °C.
6. Select a single colony and culture overnight at 37 °C in a liquid LB medium, and isolate the plasmid using the QIAprep Spin Miniprep Kit.
7. Sequence the insert DNA using primers in the vector backbone (*see* **Note 11**).

3.6 Preparation of RNA-Targeting Aptamer (RT-Aptamer)

The specific RNA-recognition aptamer is prepared as shown in Fig. 2.

1. The stem regions are removed from the quencher-binding stem-loop structures [20, 22], leading to the formation of the flexible quencher-binding loops that lack in quencher-binding activities.

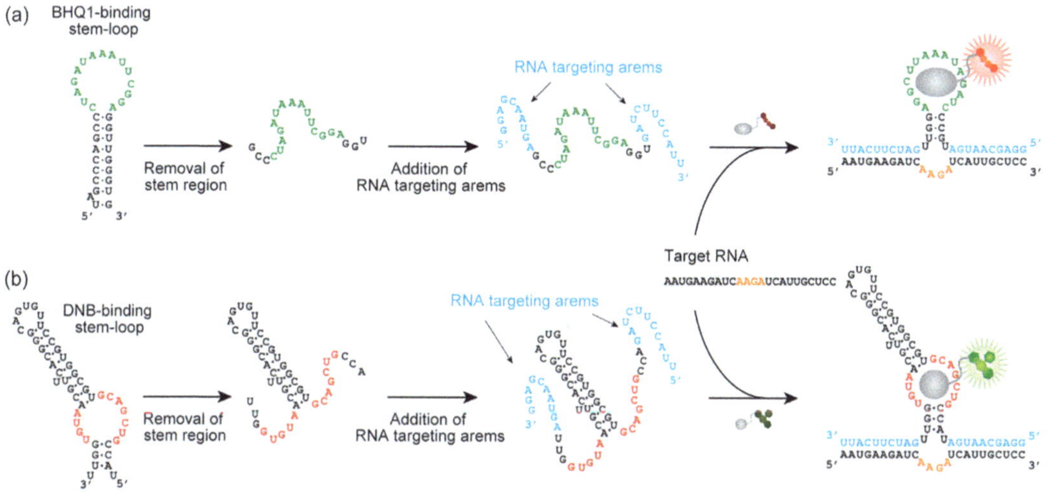

Fig. 2 Design of RNA-targeting aptamer: (**a**) BHQ1-binding aptamer, (**b**) DNB-binding aptamer

2. Two short RNA sequences (RNA targeting arms), complementary to a 24-base oligonucleotide in a target RNA sequence, are attached to the flexible structures (*see* **Note 12**).
3. The RNA-targeting arms hybridize to a target RNA, consequently forming the quencher-binding loop on the target mRNA (*see* **Note 13**).

3.7 Fluorescence In Situ Hybridization (FISH) in Fixed Cells with the RT-Aptamer [27]

1. Fix HeLa cells on 96-well plates (3.0×10^3 cells per well) with a 4% paraformaldehyde at room temperature for 15 min.
2. Wash the cells twice with PBS.
3. Permeabilize the cells with 70% ethanol overnight at 4 °C.
4. Soak the cells in 2× SSC buffer at 50 °C for 15 min.
5. Incubate the cells at 37 °C for 24 h with the pre-hybridization buffer.
6. Incubate the cells with 100 ng of the RT-aptamers in 2× SSC buffer (50 μL) for 2 h at 37 °C.
7. Treat the cells with 30 μL of 1 μg/mL Hoechst 33342 in 2× SSC buffer at 37 °C for 30 min.
8. Add a 30 μL of 5 μM BHQ1-Cy5 and DNB-DY probes in 2× SSC buffer, incubate at 37 °C for 30 min, and then wash out excess probe with 2× SSC buffer.
9. Observe the target RNAs under a confocal microscope (*see* **Note 14**).

3.8 Preparation of an Expression Vector Encoding RT-Aptamer [28–30]

1. Prepare the dsDNA template encoding RT-aptamer by annealing two 100 μM synthetic ssDNA oligos, RT aptamer templates 1 and 2 (Fig. 3), in annealing buffer.
2. Prepare the ligation mixture using *Bgl* II- *Hind* III linearized pSuper.neo plasmid as a vector, and incubate the mixture for 30 min at 16 °C to allow the ligation reaction.
3. Transform DH5α with the ligation product, and heat shock at 42 °C for 1 min, and spread the cells on an LB plate containing 100 μg/mL of carbenicillin.
4. Incubate the plate overnight at 37 °C.
5. Select a single colony, and culture in a liquid LB medium overnight at 37 °C, and isolate the plasmid using a HiSpeed Plasmid Midi Kit.
6. Sequence the insert DNAs using a universal primer (*see* **Note 11**).

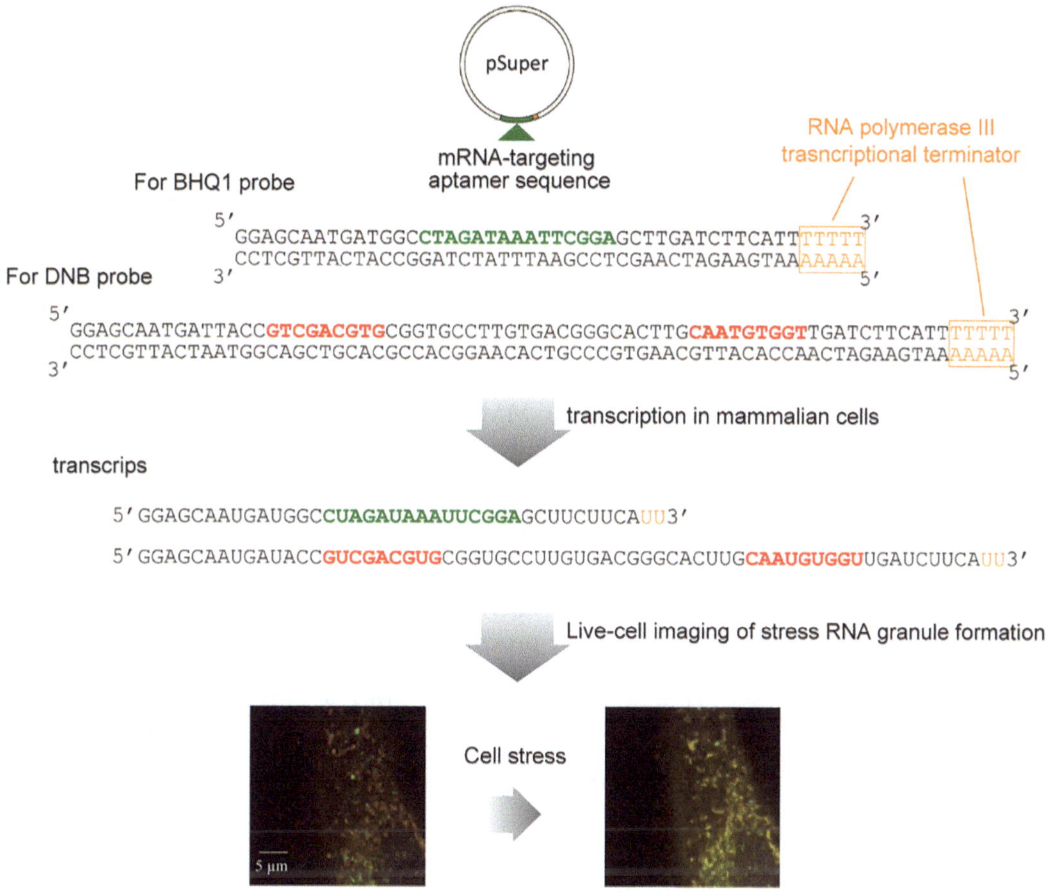

Fig. 3 Observation of RNA granule formation in living cells

3.9 Live-Cell Imaging of Multiple Endogenous mRNAs Targets with RT-Aptamers

1. Seed HeLa cells on 12-well plates (1.0×10^5 cells per well), and culture overnight at 37 °C.
2. Transfect the cells with two kinds of pSuper.neo vectors encoding an RT-BHQ1 or an RT-DNB aptamer by using FuGENE HD transfection reagent, according to the manufacturer's protocol [31].
3. Detach the cells from the plates with 0.25 w/v % trypsin, and re-seed the cells on 96-well plates (5.0×10^3 cells per well) in the complete growth medium.
4. Culture the cells for 4 h at 37 °C in a humidified 5% CO_2 incubator.
5. Stain the cells with the BHQ1-Cy5 and DNB-DY probes (final concentration, 5 μM each) and Hoechst 33342 (final concentration, 1 μg/mL) in the complete growth medium (50 μL) for 5 min at 37 °C in a humidified 5% CO_2 incubator (*see* **Notes 15** and **16**).

6. Wash the cells with 1× PBS, and then induce the cell stress with 500 μM $NaAsO_2$.
7. Observe the stress mRNA granule formation under a confocal microscope (Fig. 3) (*see* **Note 14**).

4 Notes

1. A 5 w/v % solution in DMSO is prepared before use.
2. $deSO_3$-DY520-XL was synthesized according to the previous report [32].
3. We used DMT-MM as a coupling reagent, but other water-soluble coupling reagents can be used.
4. We monitored the reaction progress with TLC.
5. We monitored the reaction progress with LC-MS.
6. The yield is normally expected to be around 80% or higher. If the yield is considerably lower than this value, we recommend that the whole process should be repeated with new reagents, especially the Cy5-NHS-ester and $deSO_3$-DY520-XL stock solution, methoxytrityl-N-dPEG8-TFP ester, and HATU.
7. In the preparation of the first DNA pool, we used a sufficient amount of DNA template to maintain the diversity of the library.
8. A NAP-5 column is equilibrated with three column volumes of RNase-free water.
9. The percentage of RNA bound to the resin is calculated as 100 (AbsT−AbsF)/AbsT, where AbsT is the UV absorbance of the RNA solution before binding to the resin and AbsF is that of the flow-through fraction.
10. PCR was stopped during the positive acceleration phase to avoid non-specific amplification. The purity of PCR products was assessed by 8 w/v % native polyacrylamide gel electrophoresis.
11. M13-RV was used as a universal primer for sequencing.
12. The RNA targeting arms are connected with a 3-base-pair stem segment through U mononucleotide bridges.
13. We normally single out an effective siRNA site as an RT-aptamer recognition sequence. Such a site might exhibit high levels of exposure for hybridization.
14. Cell images were taken with a vertical range of 4 μm and displayed as maximum intensity projection (MIP). Time-lapse images were taken at 20–30 s intervals for up to 10 min. The cell images were taken with a vertical range of 2 μm, and each image stack was then projected onto a single plane. In Fig. 3,

two kinds of RT-aptamers targeted b-actin and CTTN mRNAs, respectively.

15. We added the BHQ1-Cy5 and DNB-DY probes from stock solutions in DMSO (1 mM) and Hoechst 33342 from a stock solution in H_2O (1 mg/mL) to wells filled with media.
16. Overstaining (> 5 min) results in an increase of background fluorescence.

Acknowledgments

This work was supported in part by JSPS (21 K18214, 20H02859, and 23H02082 to SS) and ZE Research Program, IAE (ZE2023B-06 to SS). We also thank M. Nakashima for manuscript preparation.

References

1. Tyagi S (2009) Imaging intracellular RNA distribution and dynamics in living cells. Nat Methods 6:331–338
2. Lampasonaa AA, Czaplinski K (2016) RNA voyeurism: a coming of age story. Methods 98:10–17
3. Park HY, Lim H, Yoon YJ, Follenzi A, Nwokafor C, Lopez-Jones M, Meng X, Singer RH (2014) Visualization of dynamics of single endogenous mRNA labeled in live mouse. Science 343:422–424
4. Santangelo PJ, Lifland AW, Curt P, Sasaki Y, Bassell GJ, Lindquist ME, Crowe JE Jr (2009) Single molecule-sensitive probes for imaging RNA in live cells. Nat Methods 6:347–349
5. Lionnet T, Czaplinski K, Darzacq X, Shav-Tal Y, Wells AL, Chao JA, Park HY, de Turris V, Lopez-Jones M, Singer RH (2011) A transgenic mouse for in vivo detection of endogenous labeled mRNA. Nat Methods 8: 165–170
6. Mili S, Moissoglu K, Macara IG (2008) Genome-wide screen reveals APC-associated RNAs enriched in cell protrusions. Nature 453:115–119
7. Shav-Tal Y, Darzacq X, Shenoy SM, Fusco D, Janicki SM, Spector DL, Singer RH (2004) Dynamics of single mRNPs in nuclei of living cells. Science 304:1797–1800
8. Paige JS, Nguyen-Duc T, Song W, Jaffrey SR (2012) Fluorescence imaging of cellular metabolites with RNA. Science 335:1194
9. Song W, Strack RL, Jaffrey SR (2013) Imaging bacterial protein expression using genetically encoded RNA sensors. Nat Methods 10:873–875
10. Strack RL, Disney MD, Jaffrey SR (2013) A superfolding Spinach2 reveals the dynamic nature of trinucleotide repeat-containing RNA. Nat Methods 10:1219–1224
11. Hövelmann F, Gaspar I, Loibl S, Ermilov EA, Roder B, Wengel J, Ephrussi A, Seitz O (2014) Brightness through local constraint – LNA-enhanced FIT hybridization probes for in vivo ribonucleotide particle tracking. Angew Chem Int Ed 53:11370–11375
12. Ozawa T, Natori Y, Sato M, Umezawa Y (2007) Imaging dynamics of endogenous mitochondrial RNA in single living cells. Nat Methods 4:413–419
13. Wang DO, Matsuno H, Ikeda S, Nakamura A, Yanagisawa H, Hayashi Y, Okamoto A (2012) A quick and simple FISH protocol with hybridization-sensitive fluorescent linear oligodeoxynucleotide probes. RNA 18:166–175
14. Wang DO, Okamoto A (2015) Visualization of nucleic acids with synthetic excitation-controlled fluorescent oligonucleotide probes. Methods Mol Biol 1262:166–175
15. Oomoto I, Suzuki-Hirano I, Umeshima H, Han YW, Yanagisawa H, Carlton P, Harada Y, Kengaku M, Okamoto A, Shimogori T, Wang DO (2012) ECHO-liveFISH: in vivo RNA labeling reveals dynamic regulation of nuclear RNA foci in living tissues. Nucl Acids Res 43: e126
16. Tyagi S, Kramer FR (1996) Molecular beacons: probes that fluoresce upon hybridization. Nat Biotechnol 14:303–308
17. Bratu DP, Cha BJ, Mhlanga MM, Kramer FR, Tyagi S (2003) Visualizing the distribution and

transport of mRNAs in living cells. Proc Natl Acad Sci USA 100:13308–13313

18. Sokol DL, Zhang X, Lu P, Gewirtz AM (1998) Real time detection of DNA-RNA hybridization in living cells. Proc Natl Acad Sci USA 95: 11538–11543
19. Tyagi S, Alsmadi O (2004) Imaging native beta-actin mRNA in motile fibroblasts. Biophys J 87:4153–4162
20. Murata A, Sato S, Kawazoe Y, Uesugi M (2011) Small-molecule fluorescent probes for specific RNA targets. Chem Commun 47: 4712–4714
21. Sato S, Watanabe M, Katsuda Y, Murata A, Wang DO, Uesugi M (2015) Live-cell imaging of endogenous mRNAs with a small molecule. Angew Chem Int Ed 54:1855–1858
22. Yatsuzuka K, Sato S, Pe KB, Katsuda Y, Takashima I, Watanabe M, Uesugi M (2018) Live-cell imaging of multiple endogenous mRNAs permits the direct observation of RNA granule dynamics. Chem Commun 54: 7151–7154
23. Pe KBA, Yatsuzuka K, Hakariya H, Kida T, Katsuda Y, Fukuda M, Sato S (2021) RNA-based cooperative protein labeling that permits direct monitoring of the intracellular concentration change of an endogenous protein. Nucleic Acids Res 49(22):e132
24. Ellington AD, Szostak JW (1990) In vitro selection of RNA molecules that bind specific ligands. Nature 346:818–822
25. Tuerk C, Gold L (1990) Systematic evolution of ligands by exponential enrichment: RNA ligands to bacteriophage T4 DNA polymerase. Science 249:505–510
26. Robertson DL, Joyce GF (1990) Selection in vitro of an RNA enzyme that specifically cleaves single-stranded DNA. Nature 344: 467–468
27. Yoshimura H, Inaguma A, Yamada T, Ozawa T (2012) Fluorescent probes for imaging endogenous β-actin mRNA in living cells using fluorescent protein-tagged Pumilio. ACS Chem Biol 7:999–1005
28. Tuschl T (2002) Expanding small RNA interference. Nat Biotechnol 20:446–448
29. Brummelkamp TR, Bernards R, Agami R (2002) A system for stable expression of short interfering RNAs in mammalian cells. Science 296:550–552
30. Brummelkamp TR, Bernards R, Agami R (2002) Stable suppression of tumorigenicity by virus-mediated RNA interference. Cancer Cell 2:243–247
31. http://www.promega.jp/resources/protocols/technical-manuals/101/fugene-hd-transfection-reagent-protocol/
32. Czerney P, Wenzel P, Schweder B, Lehmann F (2004) Compound, especially marker-dye on the basis of polymethines. U.S. Patent Appl. 20040260093

INDEX

Shin-ichi Sato (ed.), *Live-Cell RNA Imaging: Methods and Protocols*, Methods in Molecular Biology, vol. 2875, https://doi.org/10.1007/978-1-0716-4248-1,

MIX
Papier aus verantwortungsvollen Quellen
Paper from responsible sources
FSC® C105338

If you have any concerns about our products,
you can contact us on
ProductSafety@springernature.com

In case Publisher is established outside the EU,
the EU authorized representative is:
Springer Nature Customer Service Center GmbH
Europaplatz 3, 69115 Heidelberg, Germany

Printed by Libri Plureos GmbH
in Hamburg, Germany